A Photographic Atlas

for the

Zoology Laboratory

FIFTH EDITION

Kent M. Van De Graaff
Weber State University

John L. Crawley

Morton Publishing Company
925 W. Kenyon, Unit 12
Englewood, Colorado 80110

*To the naturalists, environmentalists, and
conservation biologists who are dedicated to the
preservation of animal species from extinction.*

Least chipmunk *(Tamias minimus)* is common in
shrubby areas on plateaus of Zion National Park.

Copyright 1993, 1995, 1998, 2002, 2005 by Morton Publishing Company

ISBN: 0-89582-665-8

10 9 8 7 6 5 4 3 2 1

Printed in the United States of America

Preface

Zoology is an exciting, dynamic, and challenging science. It is the study of organisms within the kingdom Animalia, and it is a fascinating discipline within the broader science of biology. Students are fortunate to be living at a time when insights and discoveries in almost all aspects of zoology are occurring at a very rapid pace. Much of the knowledge learned in a zoology course has application in improving humanity and the quality of life. An understanding of zoology is essential in establishing a secure foundation for more advanced courses in the biological sciences or health sciences.

Zoology is a visually oriented science. *A Photographic Atlas for the Zoology Laboratory* is intended to provide you with quality photographs of animals, similar to those you may have the opportunity to observe in a zoology laboratory. It is designed to accompany any zoology text or laboratory manual you may be using in the classroom. In certain courses *A Photographic Atlas for the Zoology Laboratory* could serve as the laboratory manual.

An objective of this atlas is to provide you with a balanced visual representation of the major phyla of zoological organisms. Great care has been taken to construct completely labeled, informative figures that are depicted clearly and accurately. The terms used in this atlas are in agreement with those appearing in the more commonly used college zoology texts.

Animals inhabit nearly all aquatic and terrestrial habitats of the biosphere. The greatest number of animals are marine, where the first animals probably evolved. Depending on the classification scheme, animals may be grouped into as many as 35 phyla. The most commonly known phylum is Chordata that includes the subphylum Vertebrata, or the backboned animals. Chordates, however, comprise only about 5% of all the animal species. All other animals are frequently referred to as invertebrates, and account for 95% of the animal species.

Several dissections of invertebrate and vertebrate animals were completed and photographed in the preparation of this atlas. An understanding of the structure of an animal is requisite to learning about physiological mechanisms, and even how the animal functions in its environment. The selective pressures that determine evolutionary changes frequently have an influence on anatomical structures. Studying dissected specimens, therefore, provides phylogenetic information, or how one group of organisms is related to another.

Some zoology laboratories have the resources to provide students with opportunities for doing selected invertebrate and vertebrate dissections. For these students, the photographs contained in this atlas will be a valuable source for identification of structures on your specimens as they are dissected and studied. If dissection specimens are not available, the excellent photographs of carefully dissected prepared specimens presented throughout this atlas will be an adequate substitute. Care has gone into the preparation of these specimens to depict and identify the principal body structures from representative specimens of each of the major animal phyla. As the anatomy of the various animal specimens is studied in this atlas, note the similarities of body structure from one group to another. Even the anatomy of the human organism is similar to other animals, particularly to those of other mammals.

The information contained in Chapters 1 and 2 is intended to provide you with an orientation to the basic structure of an animal and an understanding of how cells divide. This edition of *A Photographic Atlas for the Zoology Laboratory* contains a discussion of the protists contained in Chapter 3. Many zoologists regard ancestral protists to be on the evolutionary lineage of animals. The animal phyla are presented in Chapters 4 through 17. Chapter 18 of this atlas is devoted to the biology of the human animal, which is presented in many zoology textbooks and courses. In this chapter, you are provided with a complete set of photographs for each of the human body systems. Human cadavers have been carefully dissected and photographs taken to clearly depict each of the principal organs from each of the body systems. Selected radiographs (X-rays), CT scans, and MR images depict structures from living persons and thus provide an applied dimension to this portion of the atlas.

Preface to Fifth Edition

The success of the previous editions of *A Photographic Atlas for the Zoology Laboratory* provided opportunities to make changes to enhance the value of this new edition in aiding students in learning about animals. The total revision of this atlas presented in its fifth edition required an inordinate amount of planning, organization, and work. As authors we have the opportunity and obligation to listen to the critiques and suggestions from students and faculty who have used this atlas. This constructive input is appreciated and has resulted in a greatly improved atlas.

One objective in preparing this edition of the atlas was to create an inviting pedagogy. The page layout was improved by careful selection of updated, new and replacement photographs. Many illustrations were redrawn to be more appealing and accurate. Tables were redesigned to enhance access of information. Each image in this atlas was carefully evaluated for its quality, effectiveness, and accuracy. Quality photographs of detailed dissections were completed enhancing the value of this edition. The reformatting of the pedagogy, enabled more photographs, photomicrographs, enlarged images in certain chapters, and additional photographs of representative animals.

Prelude

Scientists work to determine accuracy in understanding the relationship of organisms even when it requires changing established concepts. Development, structure, function, and the fossil record and geological dating are used to establish systematics and classify organisms. As new techniques become available, they too aid in our understanding of evolutionary relationships between groups of organisms and closely related species.

In 1758 Carolus Linnaeus, a Swedish naturalist, assigned all known kinds of organisms into two kingdoms–plants and animals. For over two centuries, this dichotomy of plants and animals served biologists well. In 1969, Robert H. Whittaker convincingly made a case for a five-kingdom system comprised of Monera, Protista, Fungi, Plantae, and Animalia. The five-kingdom basis of systematics prevailed for over twenty years, and is now being challenged with a new system that includes three domains (superkingdoms) and four kingdoms (see exhibit 1). This new system, which is used in this edition of *A Photographic Atlas for the Zoology Laboratory*, is based on criteria used in the past and new techniques in molecular biology. It is important to note, however, that a classification scheme is a human construct subject to alteration as additional knowledge is obtained.

Exhibit I Domains, Kingdoms, and Representative Examples

Bacteria Domain– Cyanobacteria, gram-negative and gram-positive bacteria

Archaea Domain– Methanogens, halophiles, and thermophiles

Eukarota Domain– Eukaryotes, single-celled and multicelled organisms; fungi, protists, plants, and animals

Oscillatoria, a cyanobacterium that reproduces through fragmentation

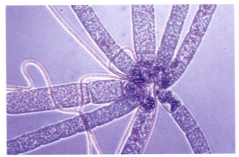

Thiothrix, a thermophile that oxidizes H_2S for an energy source

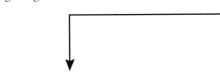

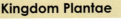

Kingdom Fungi

Kingdom Protista

Kingdom Plantae

Kingdom Animalia

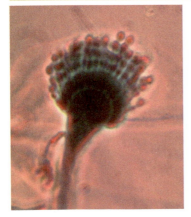

Aspergillus, a mold that reproduces asexually and sometimes sexually

Volvox, a motile green alga that reproduces asexually or sexually

Musa, the banana, is high in nutritional value and is one of the twelve most

Chamaeleo calyptratus, the veiled chameleon, is known for its ability to change colors according to its surroundings

Basic Characteristics of Domains

Domain	Characteristics
Bacteria Domain—Bacteria	Prokaryotic cell; single circular chromosome; cell wall containing peptidoglycan; chemosynthetic autotrophs, chlorophyll-based photosynthesis, photosynthetic autotrophs, and heterotrophs; gram-negative and gram-positive forms; lacking nuclear envelope; lacking organelles and cytoskeleton
Archaea Domain—Archaea	Prokaryotic cell; single circular chromosome; cell wall; membrane lipids, ribosomes, and RNA sequences; lacking nuclear envelope; some with chlorophyll-based photosynthesis; with organelles, and cytoskeleton
Eukarota Domain—Eukarya	Single-celled and multicelled organisms; nuclear envelop enclosing more than one linear chromosome; membrane-bound organelles in most; some with chlorophyll-based photosynthesis

Common Classification System of Selective Living Eukaryotes

Eukarota Domain– eukaryotes
  Kingdom Fungi– fungi
    Phylum Zygomycota– zygomycetes
    Phylum Ascomycota– ascomycetes
    Phylum Basidiomycota– basidiomycetes, teliomycetes, and ustomycetes
★ Kingdom Protista– heterotrophic and photosynthetic protists
    Phylum Myxomycota– plasmodial slime molds
    Phylum Dictyosteliomycota (Acrasiomycota)– cellular slime molds
    Phylum Oomycota– water molds
    Phylum Euglenophyta– euglenoids
    Phylum Cryptophyta– cryptomonads
    Phylum Rhodophyta– red algae
    Phylum Dinophyta (Pyrrhophyta)– dinoflagellates
    Phylum Haptophyta– haptophtes
    Phylum Bacillariophyta– diatoms (diatoms are often placed in the phylum Chrysophyta)
    Phylum Chrysophyta– chrysophytes
    Phylum Phaeophyta– brown algae
    Phylum Chlorophyta– green algae
  Kingdom Plantae– bryophytes and vascular plants
    Phylum Hepatophyta– liverworts
    Phylum Anthocerophyta– hornworts
    Phylum Bryophyta– mosses

    Phylum Psilotophyta (Psilophyta)– psilotophytes
    Phylum Lycophyta (Lycopodiophyta)– lycophytes
    Phylum Sphenophyta (Equisetophyta)– horsetails
    Phylum Pterophyta (Polypodiophyta)– ferns
    Phylum Cycadophyta– cycads
    Phylum Ginkgophyta– Ginkgo
    Phylum Coniferophyta (Pinophyta)– gymnosperms (conifers)
    Phylum Gnetophyta– gnetophytes
    Phylum Anthophyta (Magniolophyta)– angiosperms
★ Kingdom Animalia– invertebrate and vertebrate animals
    Phylum Porifera- sponges
    Phylum Cnidaria- coral, hydra, and jellyfish
    Phylum Ctenophora- comb jellies
    Phylum Platyhelminthes- flatworms
    Phylum Rotifera- rotifers
    Phylum Bryozoa
    Phylum Brachiopoda
    Phylum Phoronida
    Phylum Nemertea- proboscis worms
    Phylum Mollusca- clams, snails, and squids
    Phylum Annelida- segmented worms
    Phylum Nematoda- round worms
    Phylum Arthropoda- crustaceans, insects, and spiders
    Phylum Echinodermata- sea stars and sea urchines
    Phylum Chordata- lancelets, tunicates, and vertebrates

★ Only selective information of the Kingdoms Protista and Animalia is presented in this atlas.

Some Representatives of the Kingdom Animalia

Phylum and Representative Kinds	Characteristics
Porifera: sponges	Multicellular, aquatic animals, with stiff skeletons and bodies perforated by pores
Cnidaria: corals, hydra, and jellyfish	Aquatic animals, radially symmetrical, mouth surrounded by tentacles bearing cnidocytes (stinging cells); body composed of epidermis and gastrodermis, separated by mesoglea
Platyhelminthes: flatworms	Elongated, flattened, and bilaterally symmetrical; distinct head containing ganglia; nerve cords; protonephridia or flame cells
Mollusca: clams, snails, and squids	Bilaterally symmetrical with a true coelom, containing a mantle; many have muscular foot and protective shell
Annelida: segmented worms	Body segmented (except leeches); a series of hearts; hydrostatic skeleton and circular and longitudinal muscles
Nematoda: round worms	Mostly microscopic, unsegmented wormlike; body enclosed in cuticle; whip like body movement
Arthropoda: crustaceans, insects, and spiders	Body segmented; paired and jointed appendages; chitinous exoskeleton; hemocoel for blood flow
Echinodermata: sea stars and sea urchines	Larvae have bilateral symmetry; adults have pentaradial symmetry; coelom, most contain a complete digestive tract; regeneration of body partst
Chordata: lancelets, tunicates, and vertebrates	Fibrous notochord, pharyngeal gill pouches, dorsal hollow nerve cord, and postanal tail present at some stage in their life cycle

  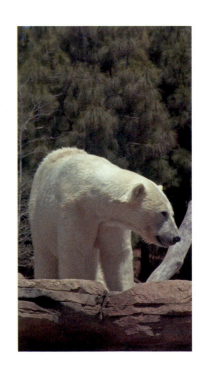

Acknowledgments

Many professionals have assisted in the preparation of *A Photographic Atlas for the Zoology Laboratory*, fifth edition, and have shared our enthusiasm of its value for students of zoology. We are especially appreciative of Drs. Wilford M. Hess, and William B. Winborn for their help in obtaining photographs and photomicrographs. Drs. Ronald A. Meyers, John F. Mull and Samuel I. Zeveloff of the Department of Zoology at Weber State University were especially helpful and supportive of this project. The radiographs, CT scans, and MR images have been made possible through the generosity of Gary M. Watts, MD and the Department of Radiology at Utah Valley Regional Medical Center. Ryan L. Van De Graaff spent many hours in assisting us with dissections and labeling. Several other students were very helpful in performing dissections. We gratefully acknowledge the assistance of Nathan A. Jacobson, Scott R. Gunn, Sandra E. Sephton, Michelle Kidder, and Michael K. Visick. A special thanks is extended to Samuel C. Mozley at North Carolina State University, P. Lynn Ruxton at Lakehead University and to Daryl E. Richter at Tulsa Community College for their proofing of the manuscript and many helpful suggestions. Thanks is extended to Christopher H. Creek and Jessica Ridd for the art throughout the book. We also appreciate Focus Design and its employees for the layout and organization of this atlas. We are indebted to Douglas Morton and the personnel at Morton Publishing Company for the opportunity, encouragement, and support to prepare this atlas.

Many of the photographs of living animals were made possible because of the cooperation and generosity of the San Diego Zoo, San Diego Wild Animal Park, Sea World (San Diego), and Hogle Zoo (Salt Lake City). We are especially appreciative to the professional zoologists at these fine institutions.

Book Team

Publisher: Doug Morton
Biology Editor: David Ferguson
Typography and Design: Focus Design
Cover: Bob Schram, Bookends Publication Design
Illustrations: Jessica Ridd

Salmon Crested or Moluccan Cockatoo (*Cacatua moluccensis*)
A native of Ceram, Saparua and Haruku in the
southern Moluccas

Table of Contents

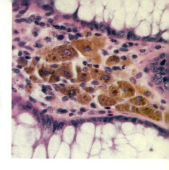

Animals are heterotrophic organisms that ingest food materials and store carbohydrate reserves as glycogen or fat. The cells of animals lack cell walls, but do contain intercellular connections including desmosomes, gap junctions, and tight junctions. Animal cells are also highly specialized into the specific kinds of tissues depicted in this chapter. Most animals are motile through the contraction of muscle fibers containing actin and myosin proteins. The complex body systems of animals include elaborate sensory and neuromotor specializations that accommodate dynamic behavioral mechanisms.

Cells are the basic structural and functional units of organization within living organisms. A cell is a minute, membrane–enclosed, protoplasmic mass consisting of chromosomes in a nucleus surrounded by cytoplasm containing the specific organelles which function independently but in coordination one with another. Based on structure, there are prokaryotic cells and eukaryotic cells.

Prokaryotic cells lack a membrane-bound nucleus, contain a single bacterial *chromosome* composed of a single strand of *nucleic acid*, contain few organelles, and have a rigid or semirigid *cell wall* outside the *cell (plasma) membrane* that provides shape to the cell. Bacteria are examples of prokaryotic, single-celled, organisms.

Eukaryotic cells contain a nucleus with multiple chromosomes, have numerous specialized *organelles*, and have a differentially permeable *cell (plasma) membrane*. Examples of eukaryotic organisms include protozoa, fungi, algae, plants, and invertebrate and vertebrate animals.

The *nucleus* is the large spheroid body within the eukaryotic cell that contains the *nucleolus, nucleoplasm,* and *chromatin*—the genetic material of the cell. The nucleus is enclosed by a double membrane called the *nuclear membrane,* or *nuclear envelope.* The nucleolus is a dense, nonmembranous body composed of protein and RNA molecules. The chromatin consists of fibers of protein and DNA molecules. Prior to cellular division, the chromatin shortens and coils into rod-shaped *chromosomes.* Chromosomes consist of DNA and proteins called *histones.*

The *cytoplasm* of the eukaryotic cell is the medium of support between the nuclear membrane and the cell membrane. *Organelles* are minute membrane-bound structures within the cytoplasm of a cell that are concerned

with specific functions. The cellular functions carried out by the organelles are referred to as *metabolism.* The functions of the principal organelles are listed in Table 1.1. In order for cells to remain alive, metabolize and maintain *homeostasis,* certain requirements must be met that include having access to nutrients and oxygen, being able to eliminate wastes, and being maintained in a constant, protective environment.

The *cell membrane* is composed of phospholipid and protein molecules, which gives form to a cell and controls the passage of material into and out of a cell. More specifically, the proteins in the cell membrane provide:

1. structural support;
2. a mechanism of molecule transport across the membrane;
3. enzymatic control of chemical reactions;
4. receptors for water–soluble hormones and other regulatory molecules;
5. cellular markers (antigens), which identify the blood and tissue type.

The phospholipids:

1. repel negative objects due to their negative charge;
2. act as receptors for fat–soluble hormones and other regulatory molecules;
3. form specific cell markers which enable like cells to attach and aggregate into tissues;
4. enter into immune reactions.

Tissues are aggregations of similar cells that perform specific functions. The tissues of the body of a multicellular animal are classified into four principal types, determined by structure and function:

1. *epithelial tissue* covers body and organ surfaces, lines body cavities and lumina (hollow portions of a body tubes), and forms various glands;
2. *connective tissue* binds, supports, and protects body parts;
3. *muscle tissue* contracts to produce movements;
4. *nervous tissue* initiates and transmits nerve impulses from one body part to another.

Table 1.1 Structure and Function of Cellular Components

Component	Structure	Function
Cell (plasma) membrane	Composed of protein and phospholipid molecules	Provides form to cell; controls passage of materials into and out of cell
Cytoplasm	Fluid to jelly-like substance	Serves as suspending medium for organelles and dissolved molecules
Endoplasmic reticulum	Interconnecting membrane-lined channels	Enables cell transport and processing of metabolic chemicals
Ribosome	Granules of nucleic acid (RNA) and protein	Synthesizes protein
Mitochondrium	Double-membraned sac with cristae (chambers)	Assembles ATP (cellular respiration)
Golgi complex	Flattened membrane-lined chambers	Synthesize carbohydrates and packages molecules for secretion
Lysosome	Membrane-surrounded sac of enzymes	Digests foreign molecules and worn cells
Centrosome	Mass of protein that may contain rod-like centrioles	Organizes spindle fibers and assists mitosis and meiosis
Vacuole	Membranous sac	Stores and excretes substances within the cytoplasm, regulates cellular turgor pressure
Microfibril and microtubule	Protein strands and tubes	Forms cytoskeleton, supports cytoplasm and transports materials
Cilium and flagellum	Cytoplasmic extensions from cell; containing microtubules	Movements of particles along cell surface or cell movement
Nucleus	Nuclear envelope (membrane), nucleolus, and chromatin (DNA)	Contains genetic code that directs cell activity; forms ribosomes

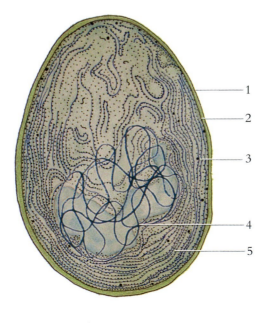

Figure 1.1 Prokaryotic cell.
1. Cell wall
2. Cell (plasma) membrane
3. Ribosomes
4. Circular molecule of DNA
5. Thyakoid membranes

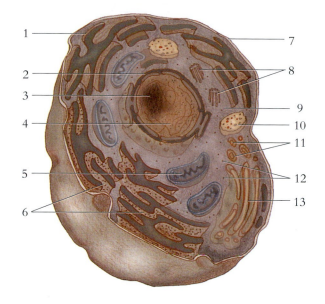

Figure 1.2 Typical animal cell.

1. Smooth endoplasmic reticulum
2. Nuclear membrane
3. Nucleolus
4. Nucleoplasm
5. Mitochondrion
6. Rough endoplasmic reticulum
7. Cell membrane
8. Centrosomes
9. Cytoplasm
10. Lysosome
11. Vesicles
12. Ribosomes
13. Golgi complex

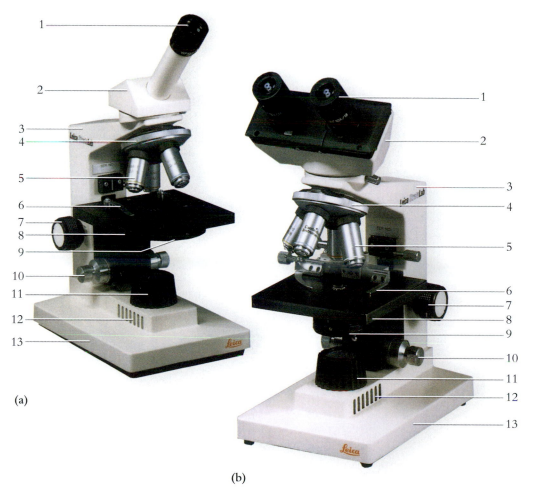

(a)

(b)

Figure 1.3 (a) Compound monocular microscope, and (b) compound binocular microscope.

1. Eyepiece (ocular)
2. Body
3. Arm
4. Nosepiece
5. Objective
6. Stage clip
7. Focus adjustment knob
8. Stage
9. Condenser
10. Fine focus adjustment knob
11. Collector lens with field diaphragm
12. Illuminator (inside)
13. Base

Photograph courtesy of Leica Inc.

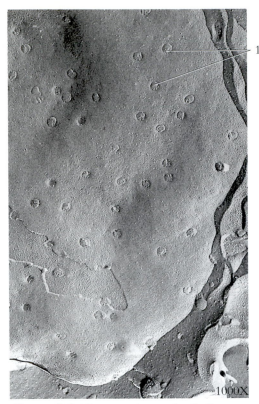

Figure 1.4 Electron micrograph of a freeze fractured nuclear envelope showing the nuclar pores.

1. Nuclear pores

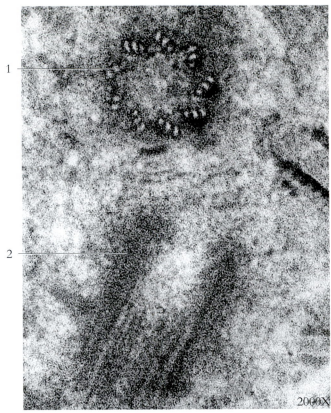

Figure 1.5 Electron micrograph of centrioles. The centrioles are positioned at right angles to one another.

1. Centriole (shown in cross section)
2. Centriole (shown in longitudinal section)

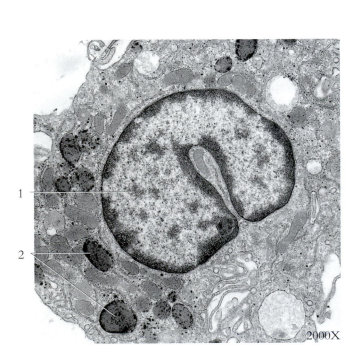

Figure 1.6 Electron micrograph of lysosomes.

1. Nucleus   2. Lysosomes

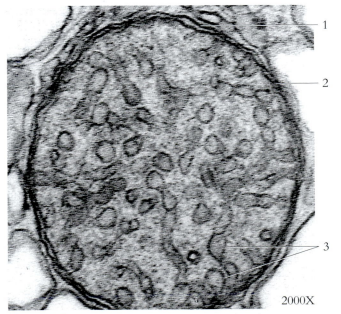

Figure 1.7 Electron micrograph of a mitochondrion.

1. Outer membrane   3. Inner membranes
2. Crista

Photographs courtesy of Scott C. Miller

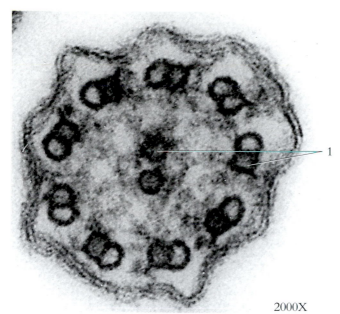

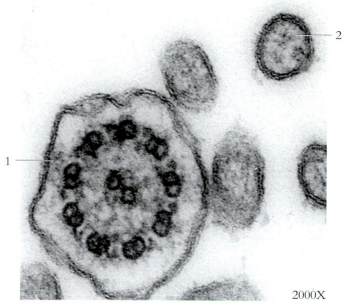

2000X

2000X

Figure 1.8 Electron micrograph of cilia (transverse section) showing the characteristic "9 + 2" arrangement of microtubules in the transverse sections.
  1. Microtubules

Figure 1.9 Electron micrograph showing the difference between a microvillus and a cilium.
  1. Microvillus   2. Cilium

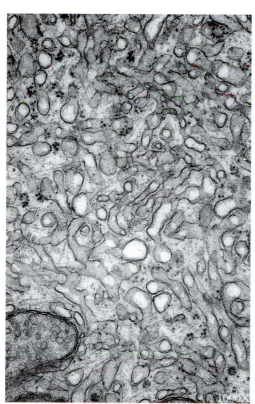

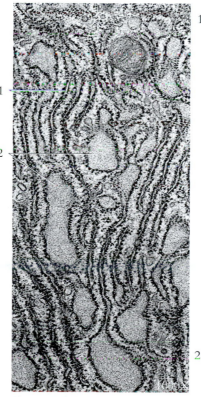

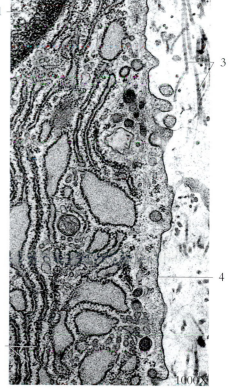

Figure 1.10 Electron micrograph of smooth endoplasmic reticulum from the testis.

Figure 1.11 Electron micrograph of rough endoplasmic reticulum.
  1. Ribosomes
  2. Cisternae

Figure 1.12 Rough endoplasmic reticulum secreting collagenous filaments to the outside of the cell.

  1. Nucleus   3. Collagenous
  2. Rough   filaments
     endoplasmic   4. Cell membrane
     reticulum

Photographs courtesy of Scott C. Miller

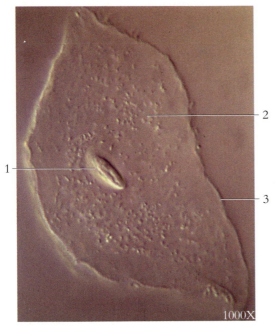

Figure 1.13 Epithelial cell from a cheek scraping.

1. Nucleus
2. Cytoplasm
3. Cell membrane

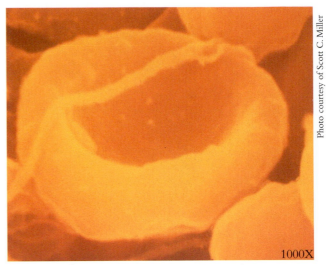

Photo courtesy of Scott C. Miller

Figure 1.14 Electron micrograph of an erythrocyte (red blood cell).

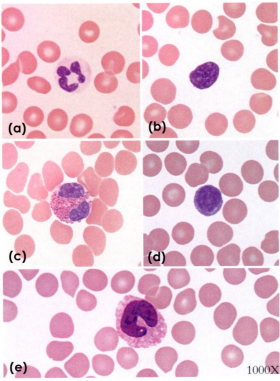

Figure 1.15 Types of leukocytes.

(a) Neutrophil   (d) Lymphocyte
(b) Basophil   (e) Monocyte
(c) Eosinophil

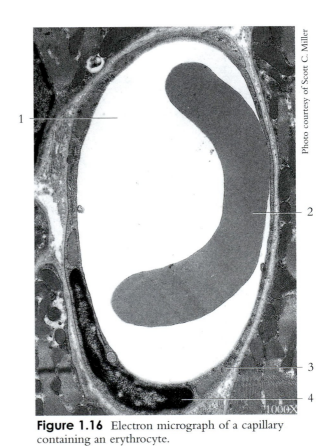

Photo courtesy of Scott C. Miller

Figure 1.16 Electron micrograph of a capillary containing an erythrocyte.

1. Lumen of capillary   3. Endothelial cell
2. Erythrocyte   4. Nucleus of endothelial cell

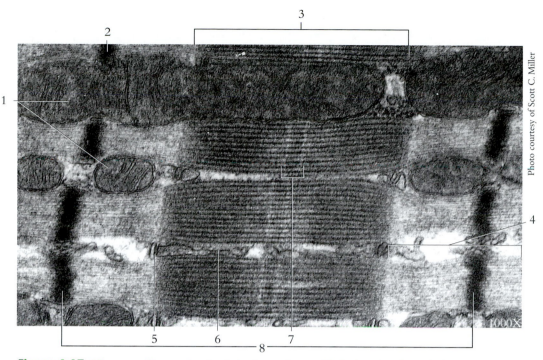

Photo courtesy of Scott C. Miller

Figure 1.17 Electron micrograph of a skeletal muscle myofibril, showing the striations.

1. Mitochondria   4. I band   7. H band
2. Z line   5. T-tubule   8. Sacromere
3. A band   6. Sarcoplasmic reticulum

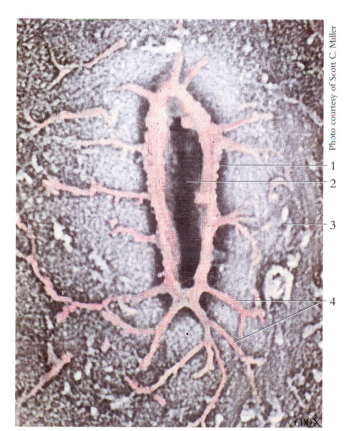

Photo courtesy of Scott C. Miller

Figure 1.18 Electron micrograph of an osteocyte (bone cell) in cortical bone matrix.

1. Lacuna   3. Bone matrix
2. Osteocyte   4. Canaliculi

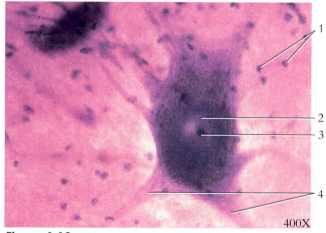

Figure 1.19 Neuron smear.
1. Nuclei of surrounding neuroglial cells
2. Nucleus of neuron
3. Nucleolus of neuron
4. Dendrites of neuron

Epithelial Tissue

Epithelial tissue covers the outside of the body and lines all organs. Its primary function is to provide protection.

Simple squamous epithelium

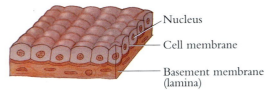

Nucleus
Cell membrane
Basement membrane (lamina)

Simple cuboidal epithelium

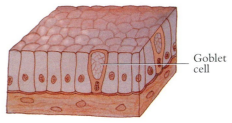

Goblet cell

Simple columnar epithelium

Connective Tissue

Connective tissue functions as a binding and supportive tissue for all other tissues in the organism.

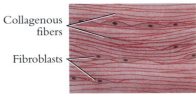

Collagenous fibers
Fibroblasts

Dense regular connective tissue

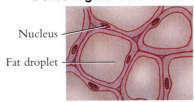

Nucleus
Fat droplet

Adipose tissue

Osteocyte
Matrix

Bone tissue

Nervous Tissue

Nervous tissue functions to receive stimuli and transmits signals from one part of the organism to another.

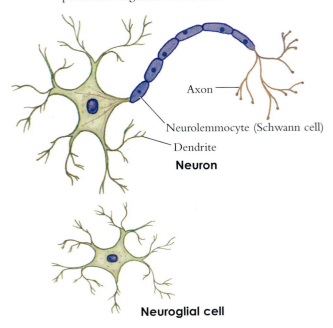

Axon
Neurolemmocyte (Schwann cell)
Dendrite

Neuron

Neuroglial cell

Muscle Tissue

Muscle tissue is a tissue adapted to contract. Muscles provide movement and functionality to the organism.

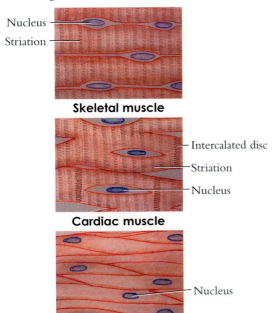

Nucleus
Striation

Skeletal muscle

Intercalated disc
Striation
Nucleus

Cardiac muscle

Nucleus

Smooth muscle

Figure 1.20 Examples of animal tissues.

Figure 1.21 Simple squamous epithelium.
1. Single layer of flattened cells

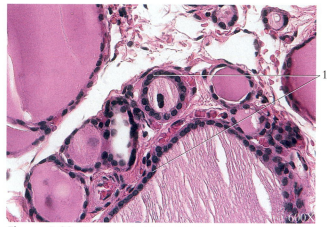

Figure 1.22 Simple cuboidal epithelium.
1. Single layer of cells with round nuclei

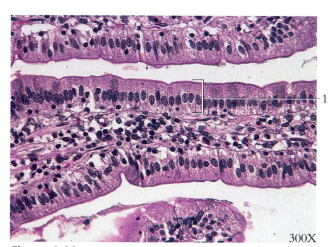

Figure 1.23 Simple columnar epithelium.
1. Single layer of cells with oval nuclei

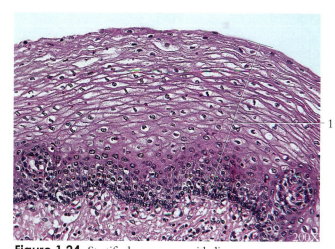

Figure 1.24 Stratified squamous epithelium.
1. Multiple layers of cells, which are flattened at the upper layer

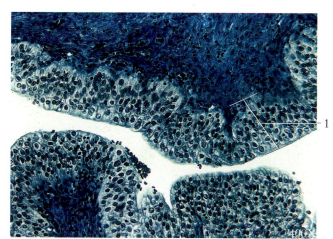

Figure 1.25 Stratified columnar epithelium.
1. Cells are balloon–like at surface

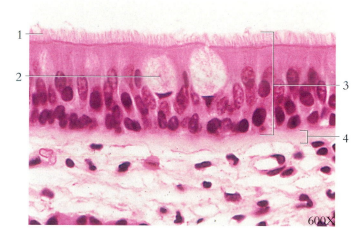

Figure 1.26 Pseudostratified columnar epithelium.
1. Cilia
2. Goblet cell
3. Pseudostratified columnar epithelium
4. Basement membrane

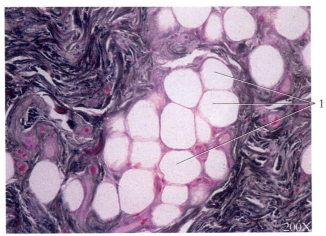

Figure 1.27 Adipose connective tissue.
1. Adipocytes (adipose cells)

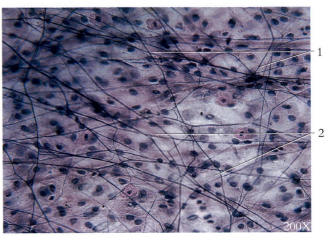

Figure 1.28 Loose connective tissue stained for fibers.
1. Elastic fibers (black)
2. Collagen fibers (pink)

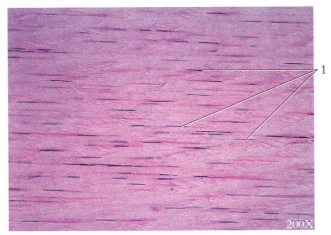

Figure 1.29 Dense regular connective tissue.
1. Nuclei of fibroblasts arranged in parallel rows

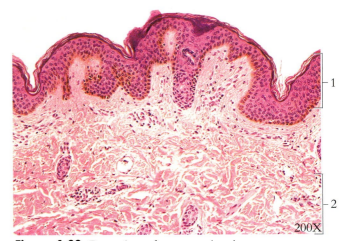

Figure 1.30 Dense irregular connective tissue.
1. Epidermis
2. Dense irregular connective tissue (reticular layer of dermis)

Figure 1.31 Electron micrograph of dense irregular connective tissue.
1. Collagenous fibers

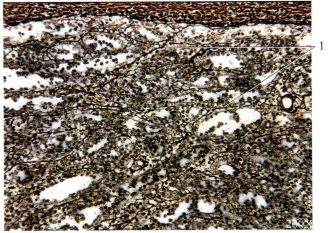

Figure 1.32 Reticular connective tissue.
1. Reticular fibers

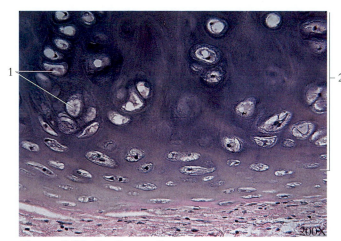

Figure 1.33 Hyaline cartilage.
1. Chondrocytes
2. Hyaline cartilage

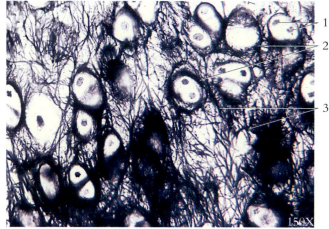

Figure 1.34 Elastic cartilage.
1. Chondrocytes   3. Elastic fibers
2. Lacunae

Figure 1.35 Fibrocartilage.
1. Chondrocytes arranged in a row

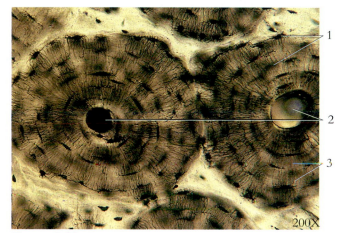

Figure 1.36 Transverse section of two osteons.
1. Lacunae   3. Lamellae
2. Central (haversian) canals

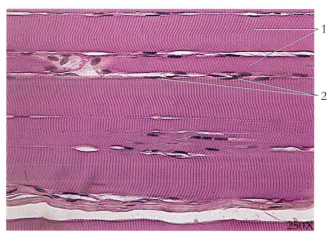

Figure 1.37 Longitudinal section of skeletal muscle tissue.
1. Skeletal muscle cells, note striations
2. Multiple nuclei in periphery of cell

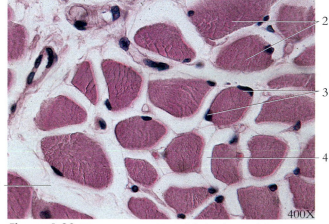

Figure 1.38 Transverse section of skeletal muscle tissue.
1. Perimysium (surrounds bundles of cells)
2. Skeletal muscle cells
3. Nuclei in periphery of cell
4. Endomysium (surrounds cells)

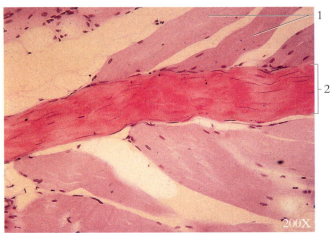

Figure 1.39 Attachment of skeletal muscle to tendon.
1. Skeletal muscle
2. Dense regular connective tissue (tendon)

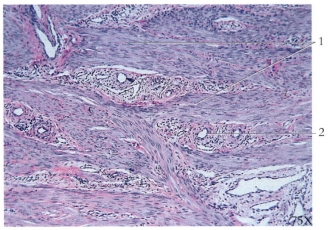

Figure 1.40 Smooth muscle tissue.
1. Smooth muscle
2. Blood vessel

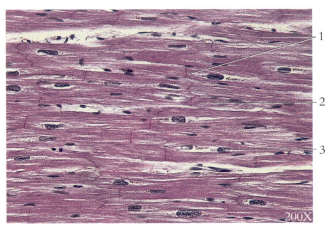

Figure 1.41 Cardiac muscle tissue.
1. Intercalated discs
2. Light-staining perinuclear sarcoplasm
3. Nucleus in center of cell

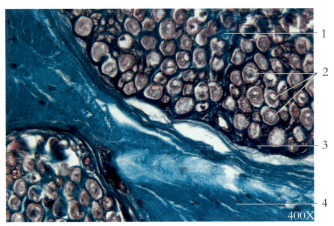

Figure 1.42 Transverse section of a nerve.
1. Endoneurium   3. Perineurium
2. Axons   4. Epineurium

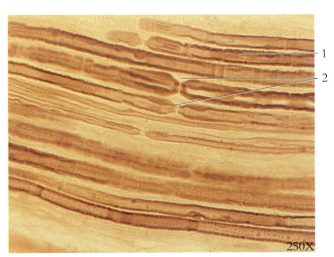

Figure 1.43 Longitudinal section of axons.
1. Myelin sheath
2. Neurofibril nodes (nodes of Ranvier)

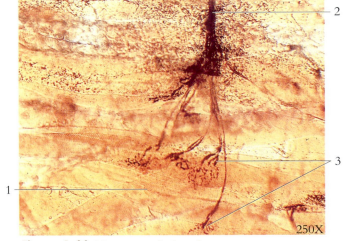

Figure 1.44 Neuromuscular junction.
1. Skeletal muscle   2. Motor nerve
   fiber   3. Motor end plates

The term *cell cycle* refers to how a multicellular organism develops, grows, and maintains and repairs body tissues. In the cell cycle, each new cell receives a complete copy of all genetic information in the parent cell, and the cytoplasmic substances and organelles to carry out hereditary instructions.

The animal cell cycle (see fig. 2.6) is divided into: 1) interphase, which includes G1, S, and G2 phases; and 2) mitosis, which includes prophase, metaphase, anaphase, and telophase. *Interphase* is the interval between successive cell divisions during which the cell is metabolizing and the chromosomes are directing RNA synthesis. The *G1 phase* is the first growth phase, the *S phase* is when DNA is replicated, and the *G2 phase* is the second growth phase. *Mitosis* (also known as karyokinesis) is the division of the nuclear parts of a cell to form two daughter nuclei with the same number of chromosomes as the original nucleus.

Like the animal cell cycle, the plant cell cycle consists of growth, synthesis, mitosis, and cytokinesis. *Growth* is the increase in cellular mass as the result of metabolism; *synthesis* is the production of DNA and RNA to regulate cellular activity; mitosis is the splitting of the nucleus and the equal separation of the chromatids; and cytokinesis is the division of the cytoplasm that accompanies mitosis.

Unlike animal cells, plant cells have a rigid cell wall that does not cleave during cytokinesis. Instead, a new cell wall is constructed between the daughter cells. Furthermore, many land plants do not have centrioles for the attachment of spindles. The microtubules in these plants form a barrel-shaped anastral spindle at each pole. Mitosis and cytokinesis in plants occurs in basically the same sequence as these processes in animal cells.

Asexual reproduction is propagation without sex; that is, the production of new individuals by processes that do not involve *gametes* (sex cells). Asexual reproduction occurs in a variety of microorganisms, fungi, plants, and animals, wherein a single parent produces offspring with characteristics identical to itself. Asexual reproduction is not dependent on the presence of other individuals. No egg or sperm is required. In asexual reproduction, all the offspring are genetically identical (except for mutants). Types of asexual reproduction include:

1. *fission*—a single cell divides to form two separate cells (bacteria, protozoans, and other one-celled organisms);
2. *sporulation*—multiple fission, many cells are formed and join together in a cystlike structure (protozoans and fungi);
3. *budding*—buds develop organisms like the parent and then detach themselves (hydras, yeast, certain plants); and
4. *fragmentation*—organisms break into two or more parts, and each part is capable of becoming a complete organism (algae, flatworms, echinoderms).

Sexual reproduction is propagation of new organisms through the union of genetic material from two parents. Sexual reproduction usually involves the fusion of haploid gametes (such as sperm and egg cells) during fertilization to form a zygote.

The major biological difference between sexual and asexual reproduction is that sexual reproduction produces genetic variation in the offspring. The combining of genetic material from the gametes produces offspring that are different from either parent and contain new combinations of characteristics. This may increase the ability of the species to survive environmental changes or to reproduce in new habitats. The only genetic variation that can arise in asexual reproduction comes from mutations.

Figure 2.1 Sexual reproduction. A pair of California quail, *Callipepla californica*, in early spring.

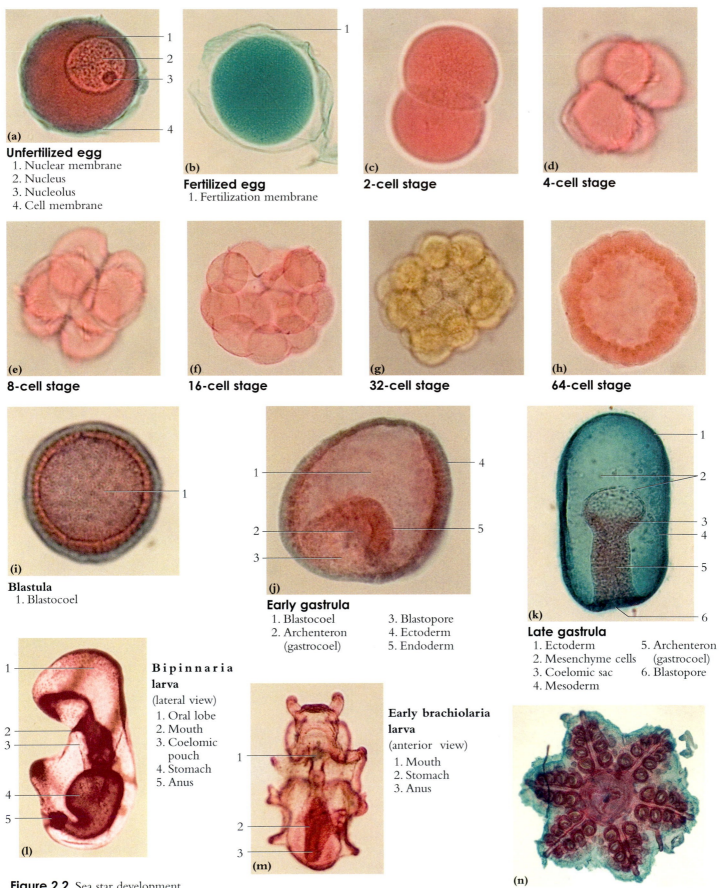

Unfertilized egg
1. Nuclear membrane
2. Nucleus
3. Nucleolus
4. Cell membrane

(b)
Fertilized egg
1. Fertilization membrane

(c)
2-cell stage

(d)
4-cell stage

(e)
8-cell stage

(f)
16-cell stage

(g)
32-cell stage

(h)
64-cell stage

(i)
Blastula
1. Blastocoel

(j)
Early gastrula
1. Blastocoel   3. Blastopore
2. Archenteron   4. Ectoderm
   (gastrocoel)   5. Endoderm

(k)
Late gastrula
1. Ectoderm   5. Archenteron
2. Mesenchyme cells   (gastrocoel)
3. Coelomic sac   6. Blastopore
4. Mesoderm

**Bipinnaria
larva**
(lateral view)
1. Oral lobe
2. Mouth
3. Coelomic
   pouch
4. Stomach
5. Anus

**Early brachiolaria
larva**
(anterior view)
1. Mouth
2. Stomach
3. Anus

(l)

(m)

(n)
Young sea star

Figure 2.2 Sea star development.

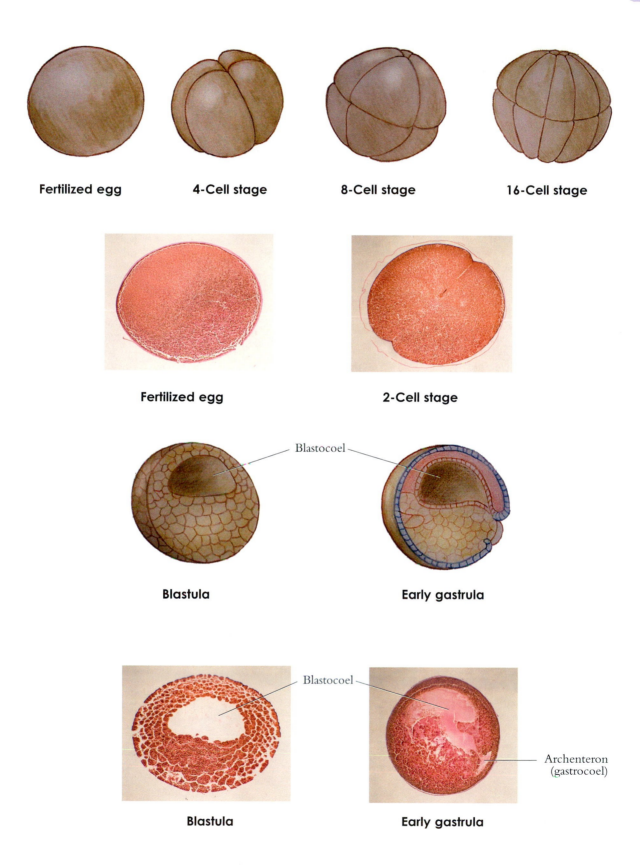

Figure 2.3 Frog development from fertilized egg to early gastrula, shown in diagram and photomicrographs.

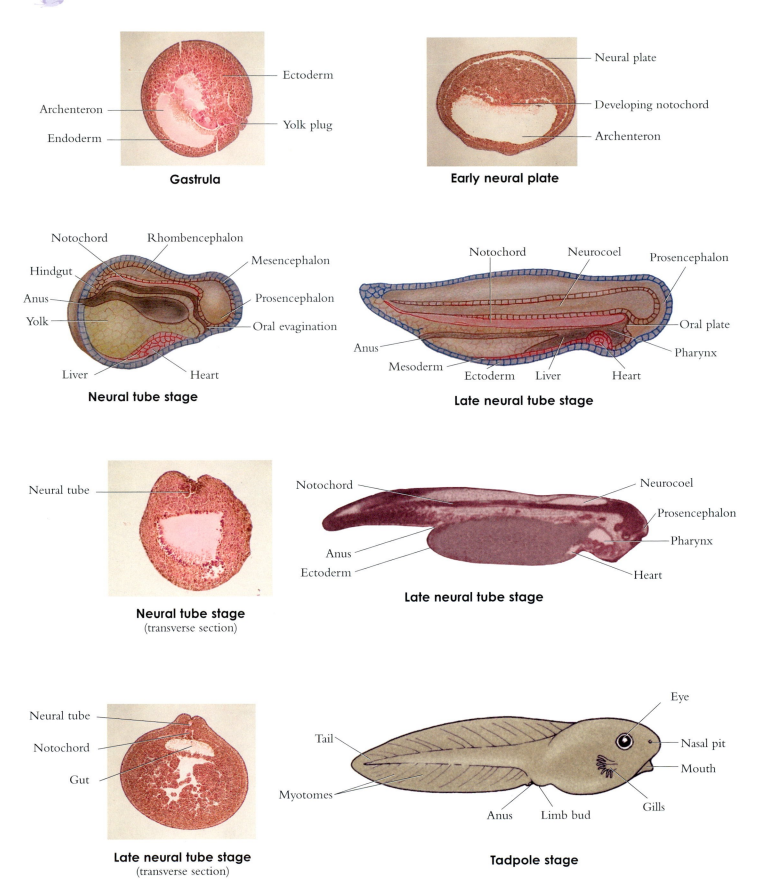

Gastrula

Early neural plate

Neural tube stage

Late neural tube stage

Neural tube stage
(transverse section)

Late neural tube stage

Late neural tube stage
(transverse section)

Tadpole stage

Figure 2.4 Frog development from gastrula to tadpole, shown in diagram and photomicrographs.

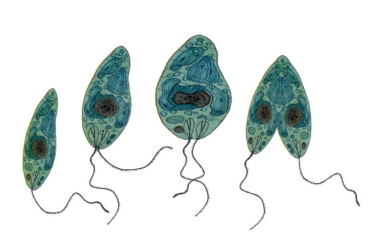

Binary fission

A single cell divides forming two separate cells. Fission occurs in bacteria, protozoans, and other single-celled organisms.

Figure 2.5 Types of asexual reproduction.

Fragmentation

An organism breaks into two or more parts, each capable of becoming a complete organism. Fragmentation occurs in flatworms and echinoderms.

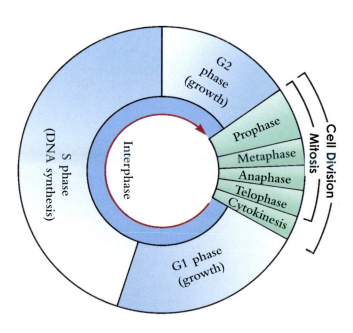

Figure 2.6 Animal cell cycle.

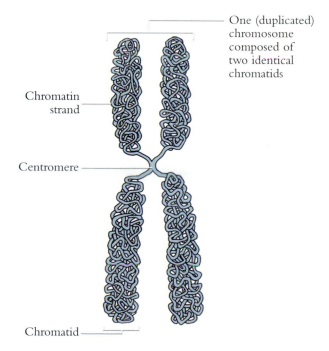

Figure 2.7 Each duplicated chromosome consists of two identical chromatids attached at the centrally located and constricted centromere.

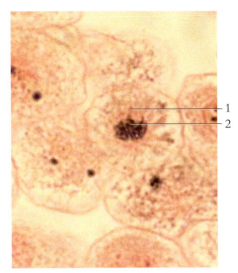

Prophase

Each chromosome consists of two chromatids jointed by a centromere. Spindle fibers extend from each centriole.
1. Aster around centriole
2. Chromosomes

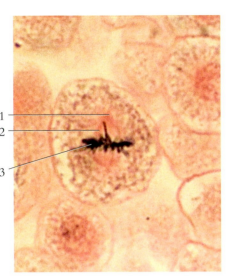

Metaphase

The chromosomes are positioned at the equator. The spindle fibers from each centriole attach to the centromeres.
1. Aster around centriole
2. Spindle fibers
3. Chromosomes at equator

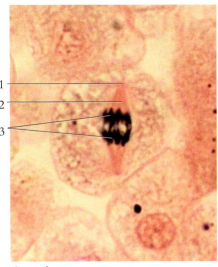

Anaphase

The centromeres split, and the sister chromatids separate as each is pulled to an opposite pole.
1. Aster around centriole
2. Spindle fibers
3. Separating chromosomes

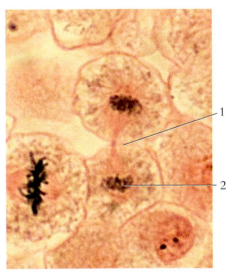

Telophase

The chromosomes lengthen and become less distinct. The cell membrane forms between the forming daughter cells.
1. New cell membrane
2. Newly forming nucleus

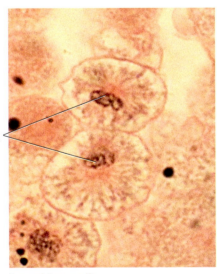

Daughter cells

The single chromosomes (former chromatids—see anaphase) continue to lengthen as the nuclear membrane reforms. Cell division is complete and the newly formed cells grow and mature.
1. Daughter nuclei

Figure 2.8 Stages of mitosis followed by cytokinesis. 500X

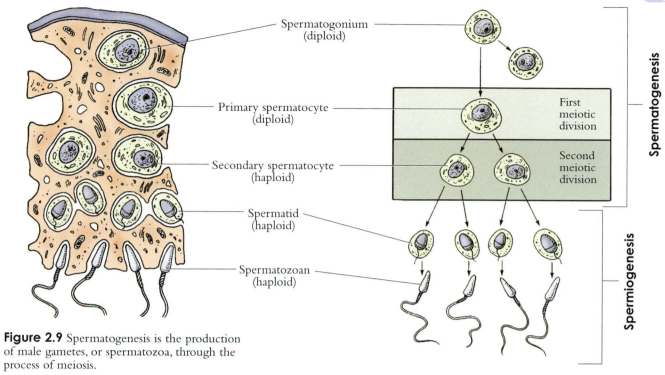

Spermatogonium
(diploid)

Primary spermatocyte
(diploid)

First
meiotic
division

Secondary spermatocyte
(haploid)

Second
meiotic
division

Spermatid
(haploid)

Spermatozoan
(haploid)

Spermatogenesis

Spermiogenesis

Figure 2.9 Spermatogenesis is the production of male gametes, or spermatozoa, through the process of meiosis.

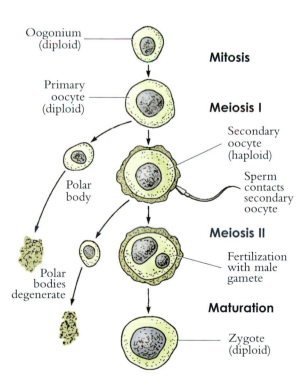

Oogonium
(diploid)

Mitosis

Primary
oocyte
(diploid)

Meiosis I

Polar
body

Secondary
oocyte
(haploid)

Sperm
contacts
secondary
oocyte

Meiosis II

Polar
bodies
degenerate

Fertilization
with male
gamete

Maturation

Zygote
(diploid)

Figure 2.10 Oogenesis is the production of female gametes, or ova, through the process of meiosis.

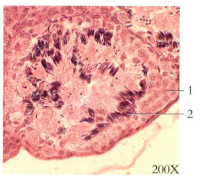

200X

Figure 2.11 Frog testis.
1. Spermatocytes
2. Developing sperm

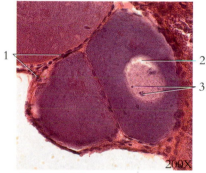

200X

Figure 2.12 Frog ovary.
1. Follicle cells
2. Geminal vesicle
3. Nucleoli

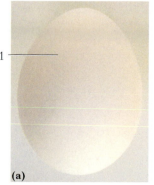

(a)

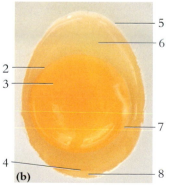

(b)

Figure 2.13 (a) Intact chicken egg and (b) a portion of the shell is removed exposing the internal structures.

1. Shell
2. Vitelline membrane
3. Yolk
4. Shell membrane

5. Shell
6. Albumen (egg white)
7. Chalaza (dense albumen)
8. Air space

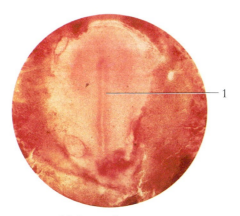

13 hour stage
1. Embryo main body formation

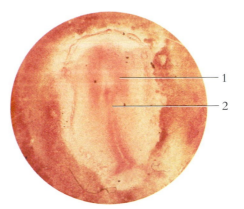

18 hour stage
1. Brain beginning to form
2. Main body

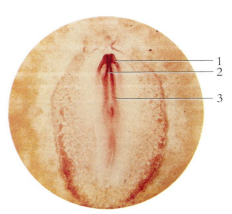

21 hour stage
1. Brain formation
2. Spinal cord beginning to form
3. Muscle plate

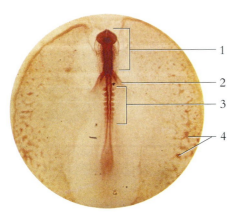

28 hour stage
1. Head fold and brain
2. Atery formation
3. Muscle plate
4. Blood vessel formation

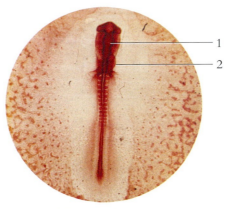

38 hour stage
1. Brain with 5 regions
2. Heart

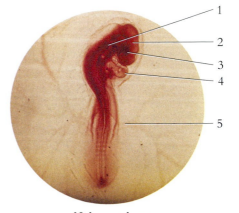

48 hour stage
1. Ear
2. Brain
3. Eye
4. Heart
5. Artery

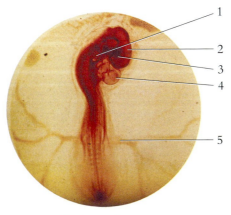

56 hour stage
1. Ear
2. Brain
3. Eye
4. Heart
5. Artery

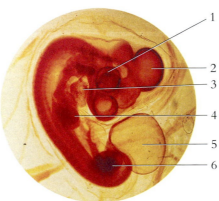

96 hour stage
1. Eye
2. Brain
3. Heart
4. Wing formation
5. Fecal sac
6. Leg formation

Figure 2.14 Stages of chick development. 20X

Protists are eukaryotic organisms that range in size from microscopic unicellular organisms to giant kelp, although some species are multicellular. Protists are all comprised of eukaryotic cells and therefore have a nucleus, mitochondria, endoplasmic reticulae, and Golgi complexes. Some contain chloroplasts. Most protists are capable of meiosis and sexual reproduction; these processes evolved a billion or more years ago and occur in nearly all complex plants and animals.

Protists are abundant in aquatic habitats, and are important constituents of plankton. Plankton are communities of organisms that drift passively or swim slowly in ponds, lakes, and oceans. Plankton are a major source of food for other aquatic organisms. Photosynthetic protists are the primary food producers in aquatic ecosystems.

The unicellular algal protists include microscopic aquatic organisms within the phyla Chrysophyta and Dinophyta. Chrysophyta are the yellow-green and golden-brown algae, and the diatoms. The cell wall of a diatom is composed largely of silica rather than cellulose. Some diatoms move in a slow, gliding way as cytoplasm glides through slits in the cell wall to propel the organism.

The Dinophyta are single-celled, algae-like organisms, the most important of which are the dinoflagellates. In most species of dinoflagellates, the cell wall is formed of armor-like plates of cellulose. Dinoflagellates are motile, having two flagella. Generally one encircles the organism in a transverse groove, and the other projects to the posterior.

Protozoa are also protists. They are small ($2\mu m$–$100\mu m$), unicellular eukaryotic organisms that lack a cell wall. Movement of protozoa is due to flagella, cilia, or pseudopodia of various sorts. In feeding upon other organisms or organic particles, they use simple diffusion, pinocytosis, active transport, or phagocytosis. Although most protozoa reproduce asexually, some species may also reproduce sexually during a portion of their life cycle. Most protozoa are harmless, although some are of immense clinical concern because they are parasitic and may cause human disease, including African sleeping sickness and malaria.

Table 3.1 Some Representatives of the Protista: Primarily Unicellular Organisms

Taxa and Representative Kinds	Characteristics
Chrysophyta—diatoms and golden algae	Diatom cell walls composed of or impregnated with silica, often with two halves; plastids often golden in Chrysophyta due to chlorophyll composition
Dinophyta—dinoflagellates	Two flagella in grooves of wall; brownish-gold plastids
Rhizopoda—amoebas	Cytoskeleton of microtubules and microfilaments; amoeboid locomotion
Apicomplexa—sporozoa and *Plasmodium*	Lack locomotor capabilities and contractile vacuoles; mostly parasitic
Sarcomastigophora—protozoa	Use flagella or pseudopodia to locomote; mostly parasitic
Euglenophyta—euglenoids	Flagellates containing chloroplasts, lacking typical cell walls
Ciliophora—ciliates and *Paramecium*	Use cilia to move and feed

Plant-like: Chrysophyta, Dinophyta

Animal-like: Rhizopoda, Apicomplexa, Sarcomastigophora, Euglenophyta, Ciliophora

Figure 3.1 Illustration of *Amoeba proteus*, a fresh-water protozoan.

Phylum Chrysophyta - diatoms and golden algae

Figure 3.2 *Biddulphia*, a colony forming colonies. These cells are beginning cell division.

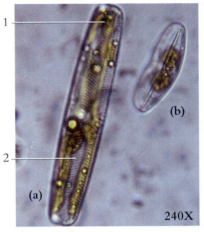

Figure 3.3 Live specimens of pennate diatoms. (a) *Navicula*, and (b) *Cymbella*.
1. Chloroplast   2. Striae

Figure 3.4 *Hyalodiscus*, a centric (radially symmetrical) diatom, from a freshwater spring in Nevada.
1. Silica cell wall   2. Chloroplasts

Figure 3.5 *Epithemia*, a distinctive (bilaterally symmetrical) freshwater diatom.

Figure 3.6 *Stephanodiscus*, a centric diatom.

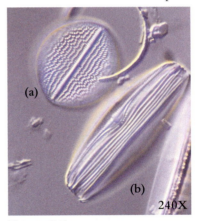

Figure 3.7 Two common freshwater diatoms. (a) *Cocconeis* and (b) *Amphora*.

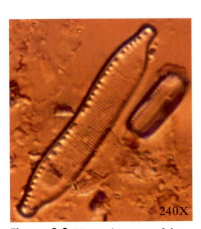

Figure 3.8 *Hantzschia*, one of the most common soil diatoms.

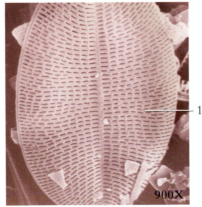

Figure 3.9 Scanning electron micrograph of *Cocconeis*, a common freshwater diatom.
1. Striae containing pores, or punctae, in the frustule (silicon cell wall).

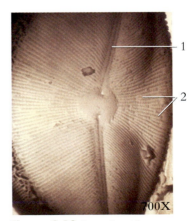

Figure 3.10 Scanning electron micrograph of the diatom *Achnanthes flexella*.
1. Raphe   2. Striae

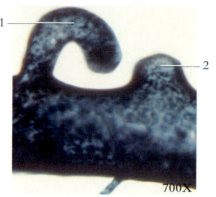

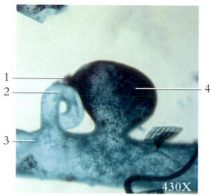

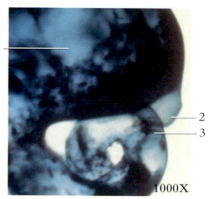

Figure 3.11 Filament with immature gametangia of the "water felt" alga, *Vaucheria*. *Vaucheria* is a chrysophyte that is widespread in fresh-water and marine habitats. It is also found in the mud of brackish areas that periodically become submerged and then exposed to air.
1. Antheridium
2. Developing oogonium

Figure 3.12 *Vaucheria*, with mature gametangia.
1. Fertilization pore
2. Antheridium
3. Chloroplasts
4. Developing oogonium

Figure 3.13 *Vaucheria*, with mature gametangia.
1. Oogonium
2. Fertilization pore
3. Antheridium

Phylum Dinophyta (Pyrrhophyta) - dinoflagellates

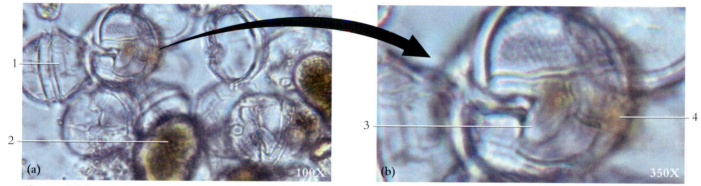

Figure 3.14 Dinoflagellates, *Peridinium*. (a) Some organisms are living; (b) others are dead and have lost their cytoplasm and consist of resistant cell walls.
1. Dead dinoflagellate
2. Living dinoflagellate
3. Cellulose plate
4. Remnant of cytoplasm

Figure 3.15 Giant clam with bluish-green coloration due to endosymbiont dinoflagellates.

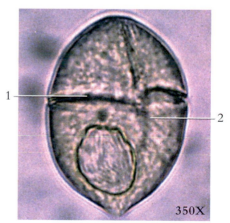

Figure 3.16 Photomicrograph of *Peridinium*. The cell wall of many dinoflagellates is composed of overlapping plates of cellulose, which are evident in this photomicrograph.
1. Transverse groove   2. Wall of cellulose plates

Phylum Rhizopoda - amoebas

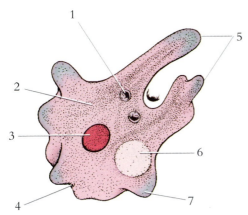

Figure 3.17 *Amoeba proteus,* is a fresh-water protozoan that moves by forming cytoplasmic extensions called pseudopodia.
1. Food vacuole   4. Cell membrane   7. Ectoplasm
2. Endoplasm   5. Pseudopodia
3. Nucleus   6. Contractile
                      vacuole

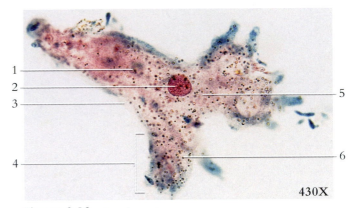

430X

Figure 3.18 *Amoeba proteus.*
1. Food vacuole   3. Cell membrane   5. Endoplasm
2. Nucleus   4. Pseudopodia   6. Ectoplasm

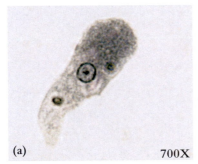
(a)   700X

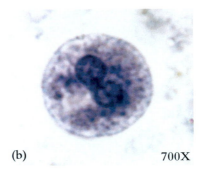
(b)   700X

Figure 3.19 Protozoan *Entamoeba histolytica,* is the causative agent of amebic dysentery, a disease most common in areas with poor sanitation. (a) A trophozoite, and (b) a cyst.

Phylum Apicomplexa - sporozoans and plasmodium

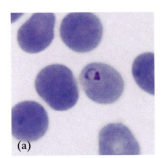
(a)

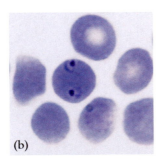
(b)

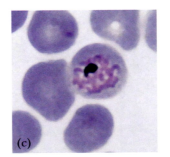
(c)

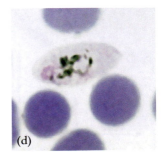
(d)

Figure 3.20 Protozoan *Plasmodium falciparum* causes malaria, which is transmitted by the female *Anopheles* mosquito. (a) The ring stage in a red blood cell, (b) a double infection, (c) a developing schizont, and (d) a gametocyte.

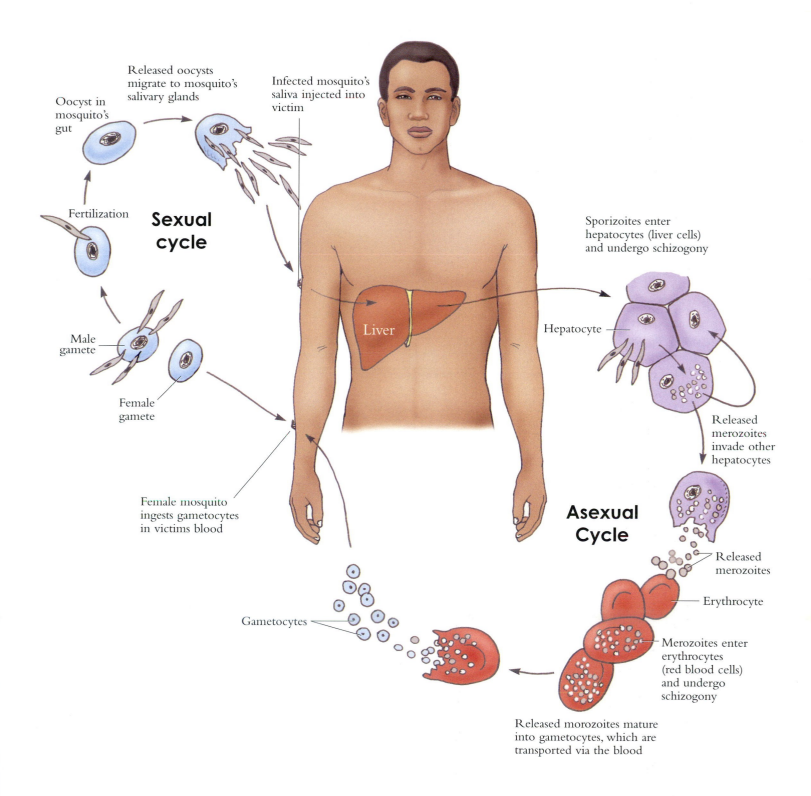

Released oocysts
migrate to mosquito's
salivary glands

Infected mosquito's
saliva injected into
victim

Oocyst in
mosquito's
gut

Fertilization

**Sexual
cycle**

Male
gamete

Female
gamete

Female mosquito
ingests gametocytes
in victims blood

Sporizoites enter
hepatocytes (liver cells)
and undergo schizogony

Hepatocyte

Liver

Released
merozoites
invade other
hepatocytes

**Asexual
Cycle**

Released
merozoites

Erythrocyte

Merozoites enter
erythrocytes
(red blood cells)
and undergo
schizogony

Released morozoites mature
into gametocytes, which are
transported via the blood

Gametocytes

Figure 3.21 Life cycle of the protozoan *Plasmodium vivax* that is carried by the female *Anopheles* mosquito, which causes malaria in humans. During the sexual cycle, sporozotes are produced in the mosquito. During the asexual cycle, merozoites are first produced in hepatocytes and then in erythrocytes.

Phylum Sarcomastiphora - flagellated protozoans

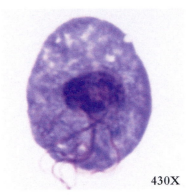

430X

Figure 3.22 Protozoan *Trichomonas vaginalis* is the causative agent of trichomoniasis. Trichomoniasis is an inflammation of the genitourinary mucosal surfaces—the urethra, vulva, vagina, and cervix in females and the urethra, prostate, and seminal vesicles in males.

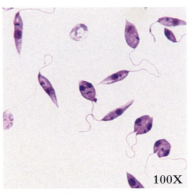

100X

Figure 3.23 Protozoan *Leishmania donovani* is the causative agent of visceral leishmaniasis, or kala-azar disease, in humans. The sandfly is the infectious host of this disease.

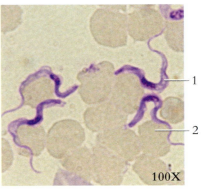

100X

Figure 3.24 Flagellated protozoan *Trypanosoma brucei* is the causative agent of African trypanosomiasis, or African sleeping sickness. The tsetse fly is the infectious host of this disease in humans.

1. *Trypanosoma brucei*
2. Red blood cell

Phylum Euglenophyta - eugenoids

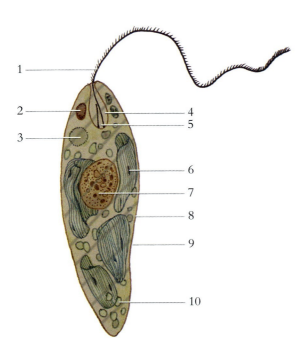

Figure 3.25 Diagram of *Euglena,* which contain flagellates that contain chloroplasts. They are freshwater organisms that have a flexible pellicle rather than a rigid cell wall.

1. Flagellum
2. Photoreceptor (eye spot)
3. Contractile vacuole
4. Reservoir
5. Basal body
6. Chloroplast
7. Nucleus
8. Pellicle
9. Cell membrane
10. Paramylum granule

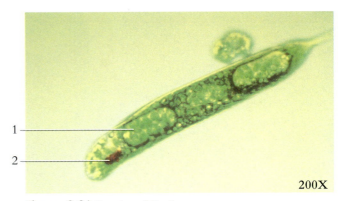

200X

Figure 3.26 Species of *Euglena*.
1. Paramylum body   2. Photoreceptor

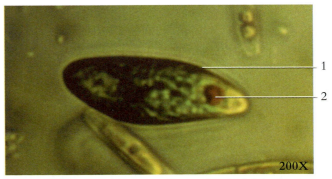

200X

Figure 3.27 Species of *Euglena* from a brackish lake in New Mexico.
1. Pellicle   2. Photoreceptor

Phylum Cillophora - ciliates and paramecia

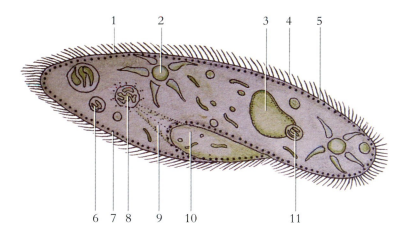

Figure 3.28 *Paramecium caudatum* is a ciliated protozoan. The poisonous trichocysts of these unicellular organisms are used for defense and capturing prey.

1. Pellicle
2. Contractile vacuole
3. Macronucleus
4. Cilia
5. Trichocyst
6. Food vacuole
7. Anal pore
8. Forming food vacuole
9. Gullet
10. Oral cavity
11. Micronucleus

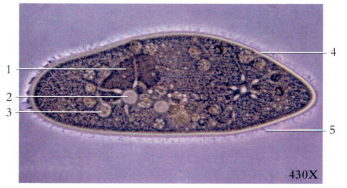

Figure 3.29 *Paramecium caudatum*, a ciliated protozoan.

1. Macronucleus
2. Contractile vacuole
3. Micronucleus
4. Pellicle
5. Cilia

Figure 3.30 *Paramecium busaria* is a unicellular, slipper-shaped organism. Paramecia are usually common in ponds containing decaying organic matter.

1. Cilia
2. Macronucleus
3. Micronucleus
4. Pellicle

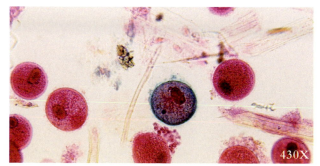

Figure 3.31 *Balantidium coli* is the causative agent of balantidiasis. Cysts in sewage-contaminated water are the infective form.

Figure 3.32 *Paramecium* in fission.

1. Contractile vacuole
2. Macronucleus
3. Micronucleus

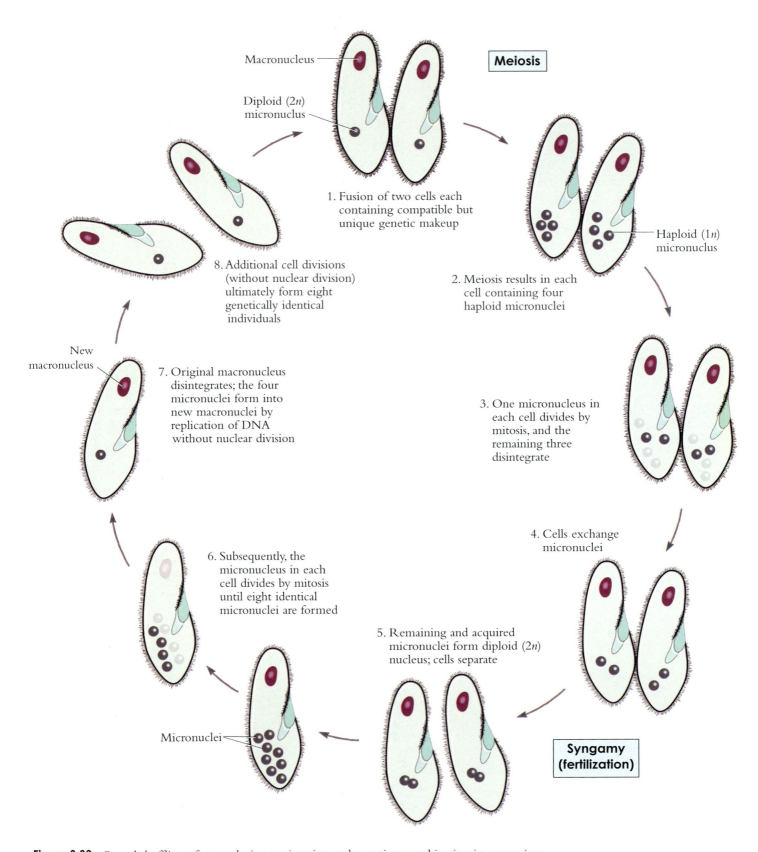

Figure 3.33   Sexual shuffling of genes during conjugation and genetic recombination in *paramecium*.

An estimated 5,000 species of sponges are contained within the phylum *Porifera*. Sponges are mostly marine organisms that lack differentiated tissues and body symmetry. The body of a sponge consists of masses of cells embedded in a supporting gelatinous matrix. The body is perforated by many pores for the passage of water.

Adult sponges are sessile, or anchored in place. Adult sponges obtain food particles through the forced circulation of water through their bodies. Water enters the central cavity, or *spongocoel*, through numerous pores, called *ostia*, and flows out of the body through the *osculum*. Water is kept moving by the action of flagellated *choanocytes*, or *collar cells*. Choanocytes obtain food particles from the water by phagocytosis. Wandering cells, called *amoebocytes*, transport nutrients from the choanocytes to other body cells. The body of a sponge is structurally supported by calcium carbonate or silica projections, called *spicules*, and by fibers of a tough protein called *spongin*.

Sponges reproduce sexually as eggs and sperm are released into the water, where fertilization occurs. The zygote develops into a free-swimming larva that soon attaches and matures into a sponge.

Sponges are a source of food for many marine animals. Sponges are harvested and prepared for commercial use by beating them to soften the spicule and spongin supporting structures, and then drying them in the sun. The soft, pliable, and absorbent nature of a prepared sponge carcass makes it ideal for wiping and cleaning.

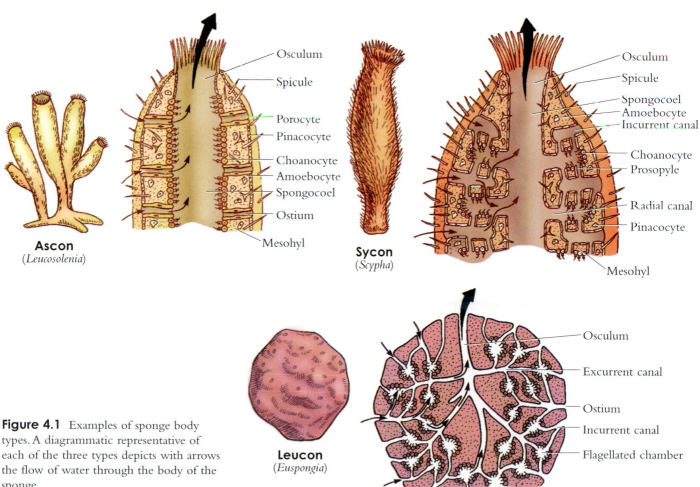

Figure 4.1 Examples of sponge body types. A diagrammatic representative of each of the three types depicts with arrows the flow of water through the body of the sponge.

Class Calcarea

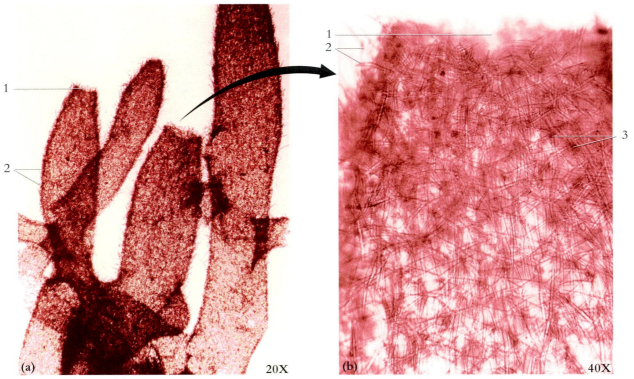

Figure 4.2 (a) *Leucosolenia* has an ascon body type. (b) High magnification of the spicules and ostia.
1. Osculum   2. Spicules   3. Ostia

Figure 4.3 (a) Sponge with an ascon body type. (b) Close up view of osculum.
1. Osculum   2. Ostia (seen from inside osculum)

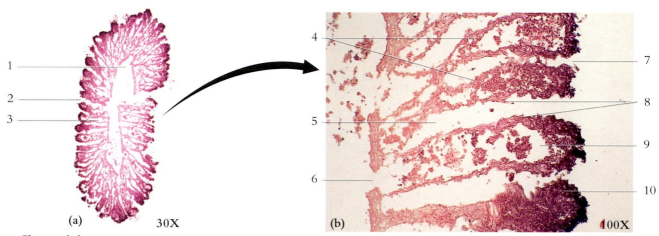

(a) 30X   (b) 100X

Figure 4.4 Transverse sections of the sponge, *Grantia* (*Scypha*). (a) Low magnification and (b) high magnification.

1. Spongocoel
2. Incurrent canal
3. Radial canal
4. Choanocytes (collar cells)
5. Incurrent canal
6. Apopyle
7. Ostium
8. Pinacocytes
9. Radial canal
10. Mesohyl

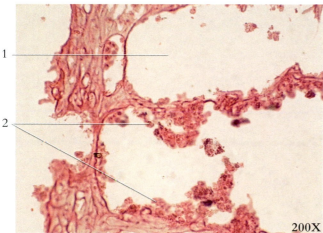

200X

Figure 4.5 Transverse section of the sponge, *Grantia* (*Scypha*), showing collar cells.

1. Radial canal
2. Choanocytes (collar cells)

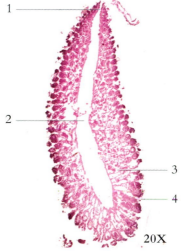

20X

Figure 4.6 Longitudinal section of the sponge, *Grantia* (*Scypha*).

1. Osculum
2. Spongocoel
3. Radial canal
4. Incurrent canal

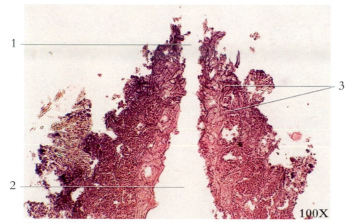

100X

Figure 4.7 Longitudinal section of the sponge, *Grantia* (*Scypha*).

1. Osculum
2. Spongocoel
3. Spicules

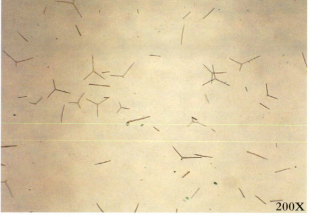

200X

Figure 4.8 Spicules of *Grantia* (*Scypha*).

Class Demospongiae

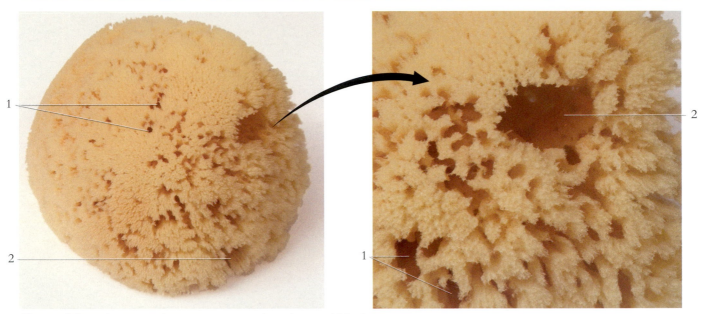

Figure 4.9 Bath sponge, class Demospongiae, has a leuconoid body structure.
1. Ostia   2. Osculum

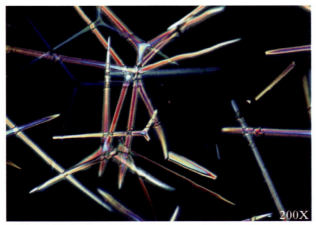

Figure 4.10 Branched silica spicules of a freshwater sponge.

200X

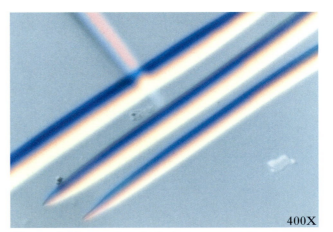

Figure 4.11 Spicules of a freshwater sponge.

400X

Figure 4.12 Leuconoid sponge growing amongst disk anemones.
1. Anemone   2. Sponge

Figure 4.13 Yellow ball sponge, *Cinachyra allocladia*.

An estimated 11,000 species are contained within the phylum *Cnidaria*. It is a large and diverse group of simply organized aquatic (mostly marine) animals that includes hydras, jellyfishes, sea anemones, and colonial corals. There are two morphological types of cnidarians: the *polyp*, or hydroid form, and the *medusa*, or jellyfish form. The polyp forms are usually sessile, or anchored in place. The medusa forms are floating or free-swimming.

Adult cnidarians have a *radially symmetrical body* that is a simple sac with the mouth opening into the *coelenteron (gastrovascular cavity)*. Food wastes are also eliminated through the mouth. The mouth is usually surrounded by tentacles bearing stinging cells called *cnidocytes*. The body is composed of an outer *epidermis* and an inner gastrodermis. A gelatinlike *mesoglea* separates these two layers.

Polyp types of cnidarians have asexual reproduction by budding. Medusa types and a few polyp types of cnidarians have sexual reproduction by gametes.

As carnivores, cnidarians are important predators within the marine food chain. They obtain their food by using their cnidocytes to sting and paralyze their prey. In tropical waters, the corals of the class Anthozoa form massive colonies. Their limy skeletons remain in place even after the polyps die and form the basis of coral reefs and some oceanic islands.

Table 5.1 Representatives of the Phylum Cnidaria

Classes and Representative Kinds	Characteristics
Hydrozoa — hydra, *Obelia*, and Portuguese man-of-war	Mainly marine; both polyp and medusa stage (polyp form only in hydra); polyp colonies in most
Scyphozoa — jellyfish	Marine coastal waters; polyp stage restricted to small larval forms
Cubozoa — box jellies	Marine coastal waters; polyp and medusa stage; square shaped when veiwed from above
Anthozoa — sea anemones, corals, and sea fans	Marine coastal waters; solitary or colonial polyps no medusa stage; partitioned gastrovascular cavity

Class Hydrozoa

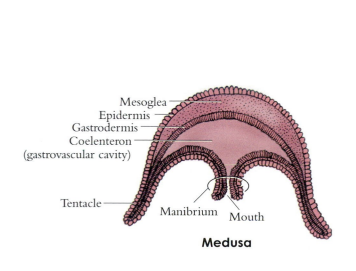

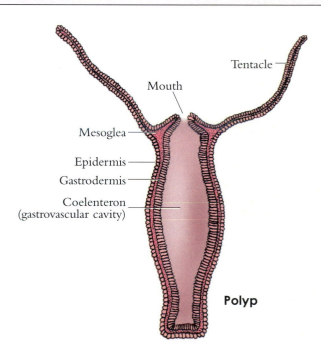

Figure 5.1 General body type of cnidarians.

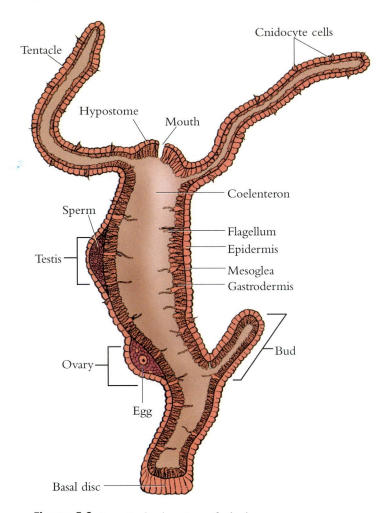

Figure 5.2 Longitudinal section of a hydra.

Tentacle
Cnidocyte cells
Hypostome
Mouth
Coelenteron
Sperm
Flagellum
Testis
Epidermis
Mesoglea
Gastrodermis
Ovary
Bud
Egg
Basal disc

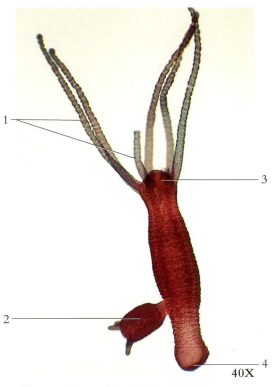

40X

Figure 5.3 Budding hydra.
1. Tentacles   3. Hypostome
2. Bud   4. Basal disc (foot)

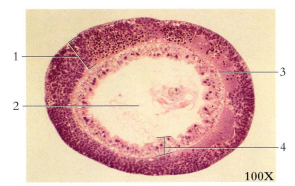

100X

Figure 5.5 Transverse section of a female hydra.
1. Epidermis (ectoderm)   4. Gastrodermis
2. Coelenteron   (endoderm)
3. Mesoglea

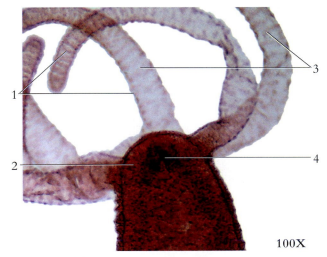

100X

Figure 5.4 Anterior end of a hydra.
1. Cnidocytes   3. Tentacles
2. Hypostome   4. Mouth

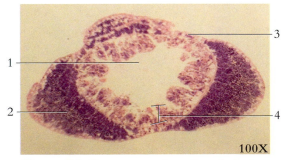

100X

Figure 5.6 Transverse section of a male hydra.
1. Coelenteron   4. Gastrodermis
2. Testes   (endoderm)
3. Epidermis (ectoderm)

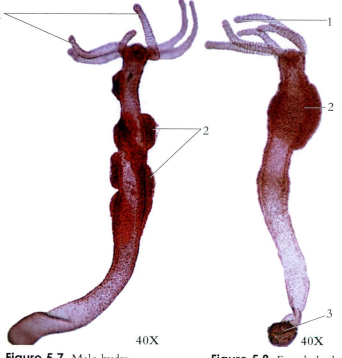

Figure 5.7 Male hydra.
1. Tentacles
2. Testes

40X

Figure 5.8 Female hydra.
1. Tentacles
2. Ovary
3. Basal disc (foot)

40X

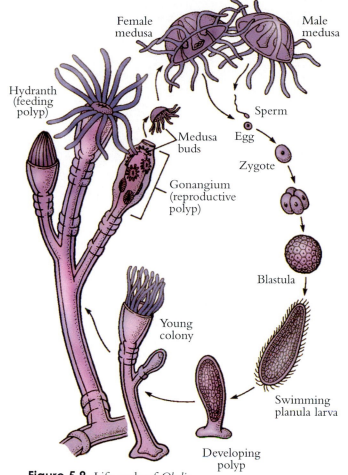

Female medusa

Male medusa

Hydranth (feeding polyp)

Sperm

Egg

Medusa buds

Zygote

Gonangium (reproductive polyp)

Blastula

Young colony

Swimming planula larva

Developing polyp

Figure 5.9 Life cycle of *Obelia*.

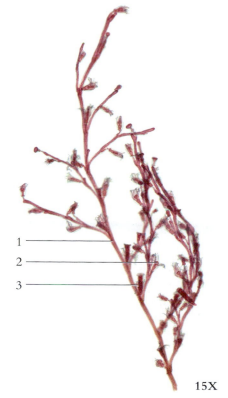

15X

Figure 5.10 *Obelia* colony.
1. Coenosarc
2. Hydranth (feeding polp)
3. Gonangium (reproductive polp)

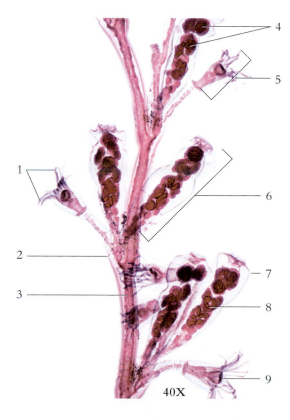

40X

Figure 5.11 Detail of *Obelia* colony.
1. Tentacles
2. Perisarc
3. Coenosarc
4. Medusa buds
5. Hydranth (feeding polp)
6. Gonangium (reproductive polp)
7. Gonotheca
8. Blastostyle
9. Hypostome

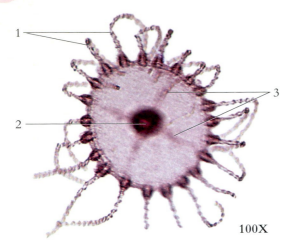

100X

Figure 5.12 *Obelia* medusa.
1. Tentacles
2. Manubrium
3. Radial canals

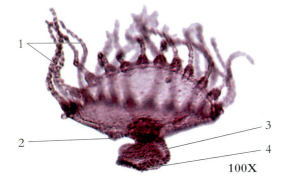

100X

Figure 5.13 *Obelia* medusa in feeding position.
1. Tentacles   3. Manubrium
2. Gonad   4. Mouth

Figure 5.14 Portuguese man-of-war, *Physalia,* is actually a colony of medusae and polyps acting as a single organism. The tentacles are comprised of three types of polyps: the gastrozooids (feeding polyps), the dactylozooids (stinging polyps), and the gonozooids (reproductive polyps).
1. Pneumatophore (float)   2. Tentacles

Class Scyphozoa

40X

Figure 5.15 *Aurelia* planula larva develops from a fertilized egg that may be retained on the oral arm of the medusa.

40X

Figure 5.16 *Aurelia* scyphistoma. The polyp is a developmental stage in the life cycle of the jellyfish.

40X

Figure 5.17 *Aurelia* strobila. Under favorable conditions, the scyphistoma develops into the strobila.
1. Developing ephyrae

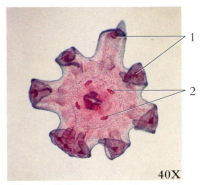

40X

Figure 5.18 *Aurelia* ephyra larva, which gradually develops into adult jellyfish.
1. Rhopalia (sense organs)
2. Gonads

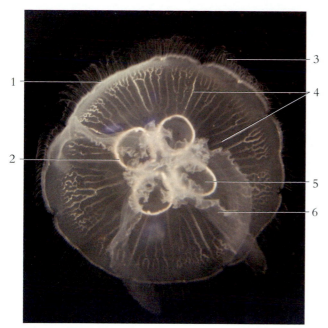

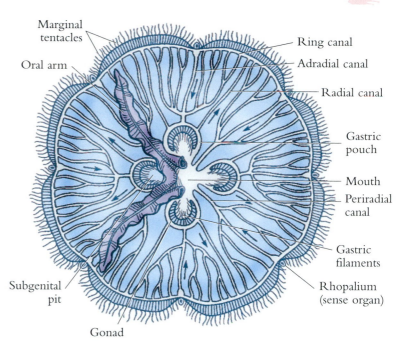

Figure 5.19 Ventral (oral) view of *Aurelia* medusa.

1. Ring canal
2. Gonad
3. Marginal tentacles
4. Radial canals
5. Subgenital pit
6. Oral arm

Figure 5.20 Ventral (oral) view of *Aurelia* medusa. In this diagram, the right oral arms have been removed. The arrows depict circulation through the canal system.

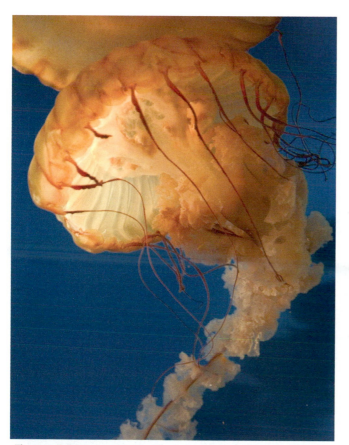

Figure 5.21 Sea nettle, *Chrysaora fuscescens*, often mass in large swarms off the Pacific coast where they feed on zooplankton.

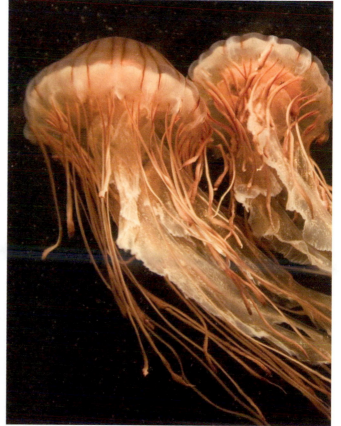

Figure 5.22 Red-striped jellyfish, *Chrysaora melanaster*, are common near the surface of the Bering sea. Mature adults have a bell close to 30 cm across and their tentacles can reach 3–6 meters.

Class Cubozoa

Figure 5.23 Box jellyfish, *Carybdea sivickisi*, are named from their cube-shaped bell. All cubozoans have four tentacles.

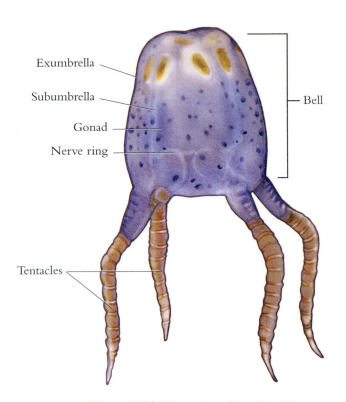

Exumbrella
Subumbrella
Gonad
Nerve ring
Bell
Tentacles

Figure 5.24 Illustration of box jellyfish, *Carybdea sivickisi*, showing basic external structures.

Class Anthozoa

Figure 5.25 Sunburst anemone, *Anthopleura sola*, gets its green coloration from symbiotic algae within it.

Figure 5.26 Two examples of anemone, disk anemones and a yellow anemone.

Figure 5.27 Disk anemones, *Actinodiscus*, form large colonies.

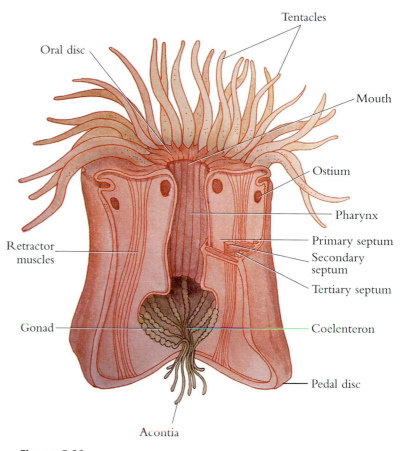

Tentacles

Oral disc

Mouth

Ostium

Pharynx

Primary septum

Secondary septum

Tertiary septum

Retractor muscles

Gonad

Coelenteron

Pedal disc

Acontia

Figure 5.28 Diagram of a partially dissected sea anemone, *Metridium*.

Figure 5.29 Brain coral, *Goniastrea*.

Figure 5.30 Skeletal structure of brain coral, *Goniastrea*.

Figure 5.31 Mushroom coral, *Rhodactis.*

Figure 5.32 Skeletal structure of mushroom coral, *Rhodactis.*

Figure 5.33 Elkshorn coral, *Acropora.*

Figure 5.34 Skeletal structure of elkshorn coral, *Acropora.*

Figure 5.35 Detailed view of the polyps of candy cane coral, *Caulastrea furcata.*

Figure 5.36 Detailed view of the polyps of glove xenia, *Xenia umbellata.*

Chapter 6
Platyhelminthes

An estimated 15,000 species of flatworms are contained within the phylum *Platyhelminthes*. Flatworms are aquatic or parasitic in the body of a host animal. They are soft-bodied, flattened animals that include planaria, flukes, and tapeworms. The *bilaterally symmetrical body* of a flatworm is composed of three tissue layers (ectoderm, mesoderm, and endoderm), and has a distinct head with a simple brain consisting of two masses of nervous tissue called *ganglia*. *Nerve cords* from the ganglia extend the length of the body. Excretory organs, called *protonephridia*, consist of *flame cells* in the body tissues and branched tubules that extend through the body and exit through pores at the body surface. The mouth opens into the *gastrovascular cavity*. The reproductive organs are well developed.

Because the human is a host animal for many flatworms, these parasitic animals are of major health concern. In tropical countries they cause high numbers of deaths, especially in children. Parasitic flatworms ingest significant quantities of nutrients, secrete toxic wastes, and generally interfere with normal physiological processes.

Table 6.1 Some Representatives of the Phylum Platyhelminthes

Classes and Representative Kinds	Characteristics
Turbellaria — planarians	Mostly free-living, carnivorous, aquatic forms; body covered by ciliated epidermis
Trematoda — flukes (schistosomes)	Parasitic with wide range of invertebrate and vertebrate hosts; suckers for attachment to host
Cestoda — tapeworms	Parasitic in many vertebrate hosts; complex life cycle with intermediate hosts; suckers or hooks on scolex for attachment to host; eggs are produced and shed within proglottids

Class Turbellaria

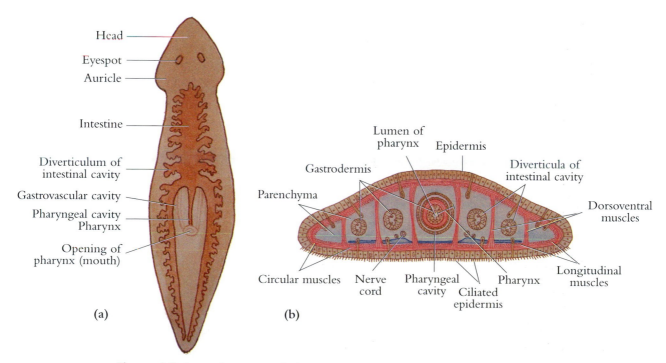

Figure 6.1 Internal anatomy of *Planaria*. (a) Longitudinal section, and (b) transverse section through the pharyngeal region.

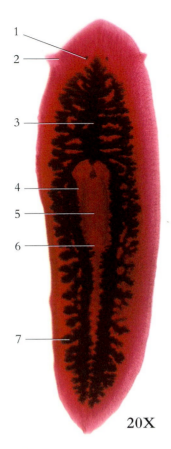

20X

Figure 6.2 *Planaria.*
1. Eyespot
2. Auricle
3. Gastrovascular cavity
4. Pharyngeal cavity
5. Pharynx
6. Opening of pharynx (mouth)
7. Diverticulum of intestinal cavity

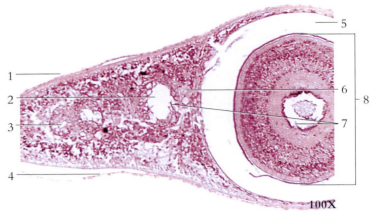

100X

Figure 6.3 Transverse section through the pharyngeal region of *Planaria.*
1. Epidermis
2. Intestinal cavity
3. Testis
4. Cilia
5. Pharyngeal cavity
6. Dorsoventral muscles
7. Gastrodermis (endoderm)
8. Pharynx

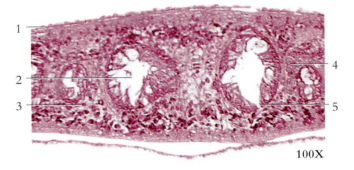

100X

Figure 6.4 Transverse section through the posterior region of *Planaria.*
1. Epidermis
2. Intestinal cavity
3. Mesenchyme
4. Dorsoventral muscles
5. Endoderm

Class Trematoda

Figure 6.5 Cow liver fluke, *Fasciola magna*, is one of the largest flukes, measuring about 3 inches.
1. Yolk gland
2. Ventral sucker
3. Oral sucker

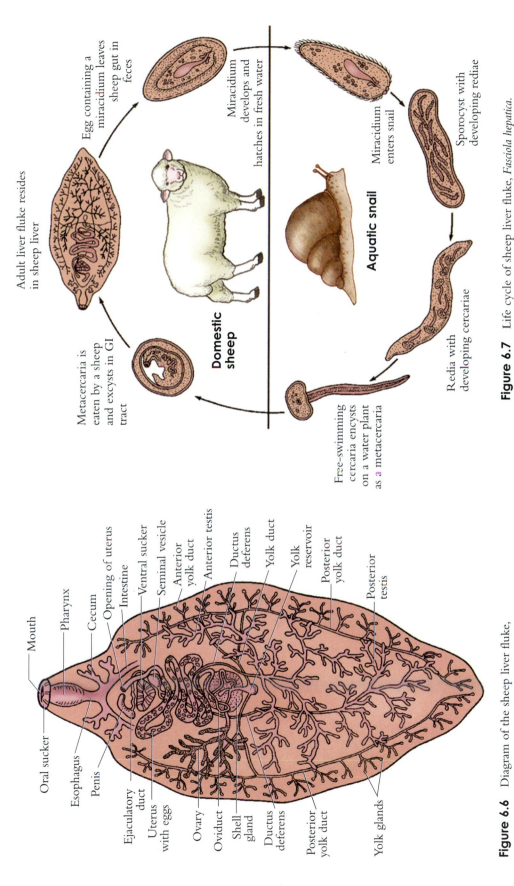

Figure 6.7 Life cycle of sheep liver fluke, *Fasciola hepatica.*

Egg containing a miracidium leaves sheep gut in feces

Adult liver fluke resides in sheep liver

Metacercaria is eaten by a sheep and excysts in GI tract

Domestic sheep

Miracidium develops and hatches in fresh water

Miracidium enters snail

Sporocyst with developing rediae

Aquatic snail

Redia with developing cercariae

Free-swimming cercaria encysts on a water plant as a metacercaria

Figure 6.6 Diagram of the sheep liver fluke, *Fasciola hepatica.*

Mouth
Pharynx
Cecum
Opening of uterus
Intestine
Ventral sucker
Seminal vesicle
Anterior yolk duct
Anterior testis
Ductus deferens
Yolk duct
Yolk reservoir
Posterior yolk duct
Posterior testis

Oral sucker
Esophagus
Penis
Ejaculatory duct
Uterus with eggs
Ovary
Oviduct
Shell gland
Ductus deferens
Posterior yolk duct
Yolk glands

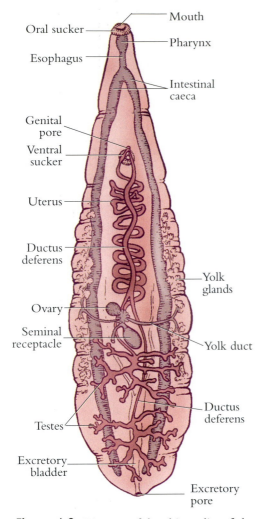

Oral sucker
Mouth
Pharynx
Esophagus
Intestinal caeca
Genital pore
Ventral sucker
Uterus
Ductus deferens
Yolk glands
Ovary
Seminal receptacle
Yolk duct
Testes
Ductus deferens
Excretory bladder
Excretory pore

Figure 6.8 Diagram of the chinese liver fluke, *Clonorchis sinensis*.

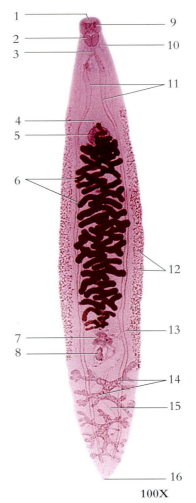

1
2
3
9
10
11
4
5
6
12
7
8
13
14
15
16

100X

Figure 6.9 Liver fluke *Clonorchis.*

1. Mouth
2. Pharynx
3. Esophagus
4. Genital pore
5. Ventral sucker
6. Uterus
7. Ovary
8. Seminal receptacle
9. Oral sucker
10. Cerebral ganglion
11. Intestine
12. Yolk glands
13. Yolk duct
14. Testis
15. Ductus (vas) deferens
16. Excretory pore

200X

Figure 6.10 Cercaria stage of a trematode species.

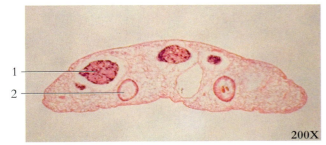

1
2

200X

Figure 6.11 Transverse section through the upper body region of *Clonorchis*.
1. Uterus
2. Intestine

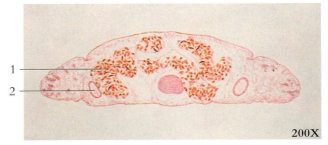

1
2

200X

Figure 6.12 Transverse section through the midbody body region of *Clonorchis*.
1. Testes
2. Intestine

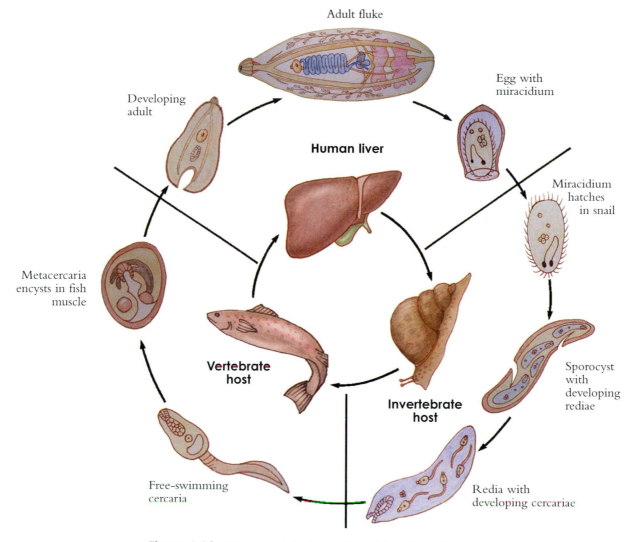

Figure 6.13 Life cycle of the human liver fluke, *Clonorchis sinesis*.

Class Cestoda

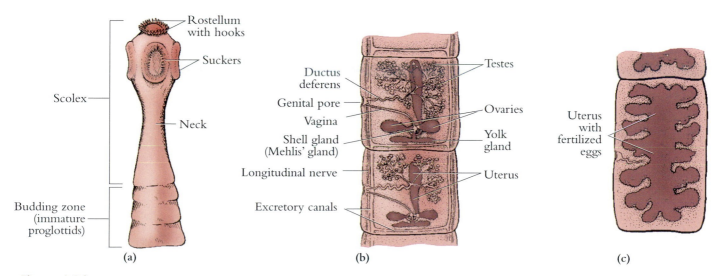

Figure 6.14 Diagrams of a parasitic tapeworm, *Taenia pisiformis*. (a) the anterior end, (b) mature proglottids, and (c) a ripe proglottid.

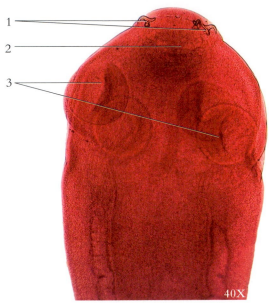

Figure 6.15   Scolex of *Taenia pisiformis*.
1. Hooks
2. Rostellum
3. Suckers

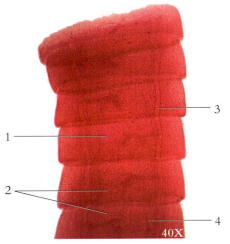

Figure 6.16   Immature proglottids
of *Taenia pisiformis*,
1. Early ovary
2. Early testes
3. Excretory canal
4. Immature vagina and ductus deferens

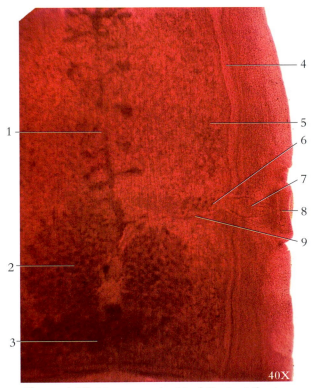

Figure 6.17 Mature proglottid of *Taenia pisiformis*.

1. Uterus	4. Excretory canal	7. Cirrus
2. Ovary	5. Testes	8. Genital pore
3. Yolk gland	6. Ductus deferens	9. Vagina

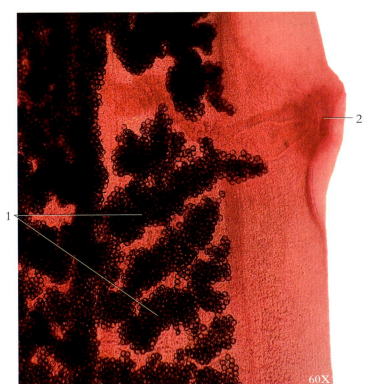

Figure 6.18 *Taenia pisiformis*, ripe proglottid.
1. Zygotes in branched uterus   2. Genital pore

An estimated 100,000 species of mollusks are contained within the phylum *Mollusca*. Mollusks are aquatic and terrestrial soft-bodied or shelled animals that include such forms as snails, clams, oysters, squids, and octopuses. They are *bilaterally symmetrical* and have true *coeloms*, and usually distinct heads. Many mollusks have a muscular foot for locomotion and a soft visceral mass enclosed by a heavy fold of tissue, called the *mantle*. Many species have a protective *shell*, which is secreted by the mantle.

Mollusca is a large and diverse phylum, with species ranging in size from a small snail that is barely macroscopic to the giant squid that is the largest invertebrate animal. Mollusks are important in marine and freshwater food chains. Some are consumed as food by humans and are of great commercial importance. These include clams, oysters, snails, mussels, squid, and octopuses. Other mollusks are hosts for disease-causing parasites and are of medical importance. These include a number of species of snails. In addition, some species of slugs and snails are of economic importance because they are devastating pests to certain crop plants.

Table 7.1 Representatives of the Phylum Mollusca

Classes and Representative Kinds	Characteristics
Polyplacophora — chitons	Marine; shell of 8 dorsal plates; broad foot
Gastropoda — snails and slugs	Marine, freshwater, and terrestrial; coiled shell; prominent head with tentacles and eyes
Bivalvia — clams, oysters, and mussels	Marine, freshwater; body compressed between two hinged shells; hatchet-shaped foot
Cephalopoda — squids and octopi	Marine; excellent swimmers, predatory; foot separated into tentacles which may contain suckers; well-developed eyes

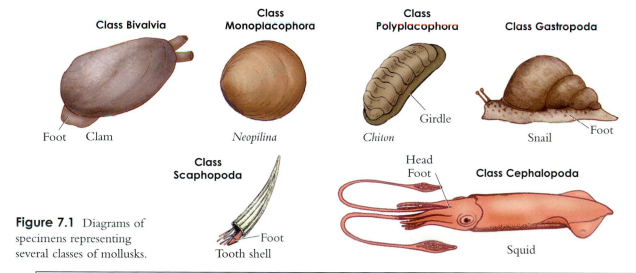

Figure 7.1 Diagrams of specimens representing several classes of mollusks.

Class Polyplacophora

Figure 7.2 Chitons are easily recognized by their eight dorsal plates.

Class Gastropoda

Figure 7.3 Snail.
1. Shell   4. Head
2. Foot   5. Tentacle
3. Eye

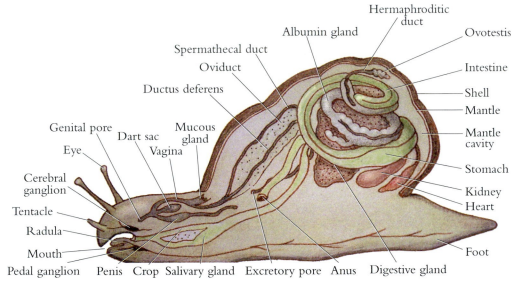

Figure 7.4 Locomotion of the slug, class Gastropoda, requires the production of mucus. Slugs differ from snails in that a shell is absent.
1. Head   3. Mantle   5. Foot
2. Tentacle   4. Mucus

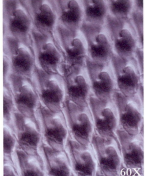

60X

Figure 7.5 Snail radule, is made up of small horny teeth made of chitin, called denticles

Figure 7.6 Diagram of pulmonate snail anatomy.

Hermaphroditic duct
Albumin gland
Spermathecal duct
Oviduct
Ovotestis
Intestine
Ductus deferens
Shell
Mantle
Genital pore
Mucous gland
Mantle cavity
Eye
Dart sac
Vagina
Stomach
Cerebral ganglion
Kidney
Heart
Tentacle
Radula
Mouth
Pedal ganglion   Penis   Crop   Salivary gland   Excretory pore   Anus   Digestive gland
Foot

Class Bivalvia

DORSAL

ANTERIOR

POSTERIOR

VENTRAL

(a)

(b)

Figure 7.7 External view of a clam shell, (a) dorsal view and (b) left valve.
1. Umbo   3. Growth lines
2. Hinge ligament

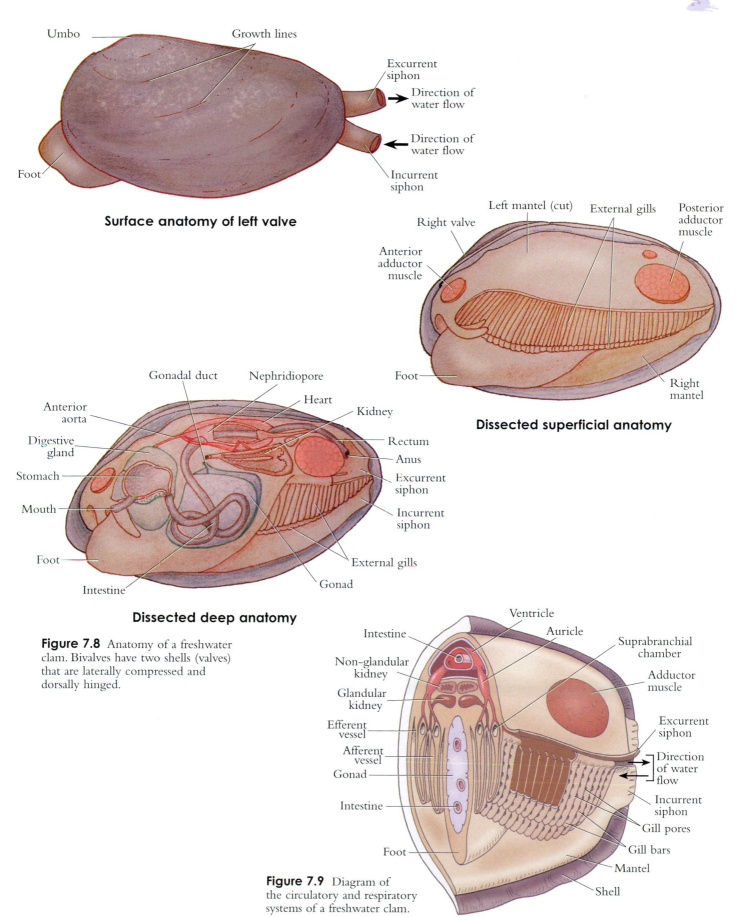

Surface anatomy of left valve

Umbo

Growth lines

Excurrent siphon

Direction of water flow

Direction of water flow

Incurrent siphon

Foot

Dissected superficial anatomy

Right valve

Left mantel (cut)

External gills

Posterior adductor muscle

Anterior adductor muscle

Foot

Right mantel

Dissected deep anatomy

Gonadal duct

Nephridiopore

Heart

Kidney

Anterior aorta

Digestive gland

Stomach

Mouth

Foot

Intestine

Rectum

Anus

Excurrent siphon

Incurrent siphon

External gills

Gonad

Figure 7.8 Anatomy of a freshwater clam. Bivalves have two shells (valves) that are laterally compressed and dorsally hinged.

Ventricle

Auricle

Suprabranchial chamber

Intestine

Non-glandular kidney

Glandular kidney

Efferent vessel

Afferent vessel

Gonad

Intestine

Foot

Adductor muscle

Excurrent siphon

Direction of water flow

Incurrent siphon

Gill pores

Gill bars

Mantel

Shell

Figure 7.9 Diagram of the circulatory and respiratory systems of a freshwater clam.

Figure 7.10 Lateral view of a clam.

1. Pericardium
2. Ventricle of heart
3. Anus
4. Posterior retractor muscle
5. Posterior adductor muscle
6. Excurrent siphon
7. Nephridium (kidney)
8. Incurrent siphon
9. Atrium of heart
10. Gills
11. Anterior retractor muscle
12. Labial palps
13. Anterior adductor muscle
14. Foot
15. Mantle

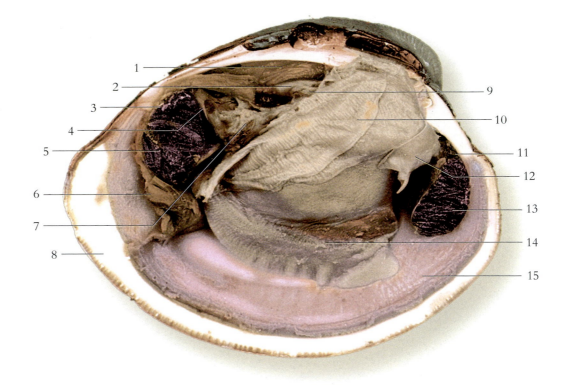

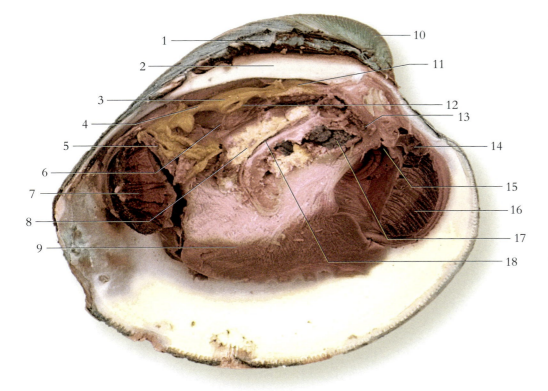

Figure 7.11 Lateral view of a clam, foot cut.

1. Hinge ligament
2. Hinge
3. Ventricle of heart
4. Posterior aorta
5. Posterior retractor muscle
6. Nephridium (kidney)
7. Posterior adductor muscle
8. Gonad
9. Foot
10. Umbo
11. Intestine
12. Opening between atrium and ventricle
13. Esophagus
14. Anterior retractor muscle
15. Mouth
16. Anterior adductor muscle
17. Digestive gland
18. Intestine

Class Cephalopoda

Figure 7.12 *Nautilus*, a cephalopod, has gas-filled chambers within its shell, as seen in the dissected specimen (b). These chambers regulate buoyancy.

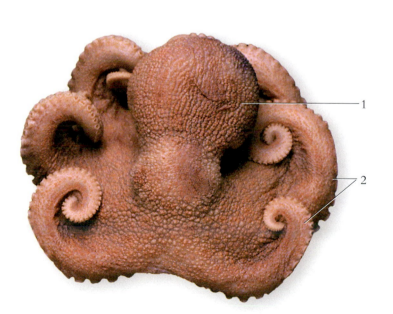

Figure 7.13 Dorsal view of an octopus collected in the Sea of Cortez, San Carlos, Mexico.
1. Body   2. Tentacles

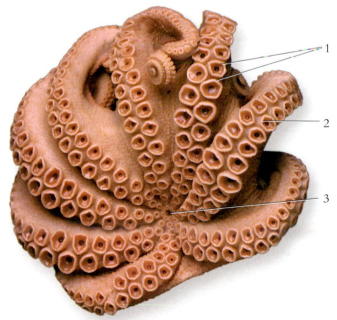

Figure 7.14 Ventral view of an octopus.
1. Suction cups   3. Mouth
2. Tentacles

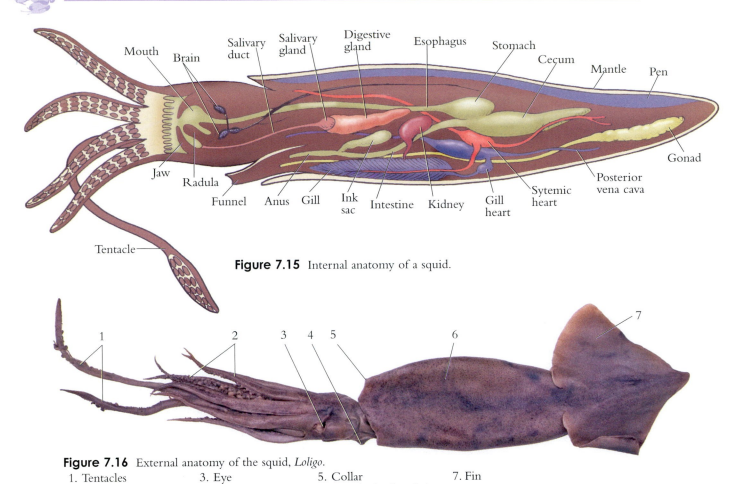

Figure 7.15 Internal anatomy of a squid.

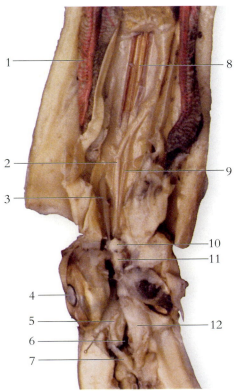

Figure 7.16 External anatomy of the squid, *Loligo*.

1. Tentacles
2. Arms
3. Eye
4. Funnel
5. Collar
6. Mantle (body tube)
7. Fin

Figure 7.18 Internal anatomy of the head region of the squid.

1. Gill
2. Esophagus
3. Pleural nerve
4. Eye
5. Radula
6. Ligula
7. Beak
8. Pen
9. Cephalic aorta
10. Visceral ganglion
11. Pedal ganglion
12. Buccal bulb

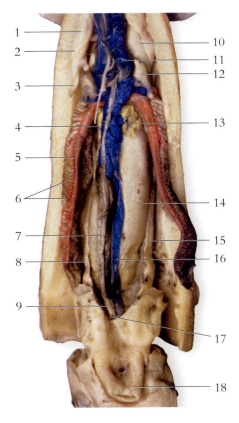

Figure 7.18 Internal anatomy of the squid.

1. Testis
2. Spermatophoric duct
3. Penis
4. Kidney
5. Branchial vein
6. Gill
7. Ink sac
8. Funnel retractor muscle
9. Rectum
10. Stomach
11. Posterior vena cava
12. Branchial heart
13. Pancreas
14. Liver
15. Funnel retractor muscle
16. Anterior vena cava
17. Anus
18. Funnel

Chapter 8
Annelida

An estimated 10,000 species of annelids are contained within the phylum *Annelida*. Annelids are marine, freshwater, and burrowing terrestrial, body-segmented worms that include such forms as sandworms, earthworms, and leeches. The annelid body contains a true coelom, a tubular digestive tract extending from the mouth to the anus, cerebral ganglia, a closed circulatory system with a series of hearts, and a hydrostatic skeleton with accompanying circular and longitudinal muscles for locomotion. Locomotion is aided in all annelids, except the leeches, by tiny chitinous bristles called *setae*. The stiff setae of burrowing annelids also aid in preventing them from being pulled out or washed out of the soil.

Sexes are separate or have both male and female organs in the same organism (*hermaphrodite*). Some of the annelids have the capability of regeneration, with each severed portion being capable of regenerating a complete organism.

Annelids are extremely important in the food chain and the general community ecology. Earthworms aerate and enrich the soil. Most of the leeches are predators, some are temporary parasites, and a few are permanent parasites.

Table 8.1 Some Representatives of the Phylum Annelida

Classes and Representative Kinds	Characteristics
Polychaeta — tubeworms and sandworms	Mostly marine; segments with parapodia
Oligochaeta — earthworms	Freshwater and burrowing terrestrial forms; small setae; poorly developed head
Hirudinea — leaches	Freshwater; some are blood-sucking parasites and others are predators; lack setae; prominent muscular suckers

Class Polychaeta

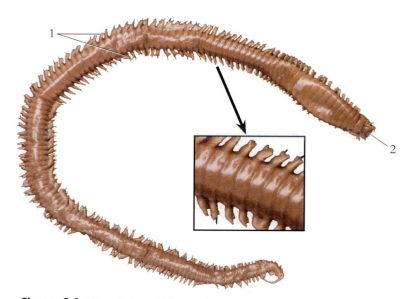

Figure 8.1 Dorsal view of the sandworm, *Neanthes*.

1. Parapodia   2. Mouth

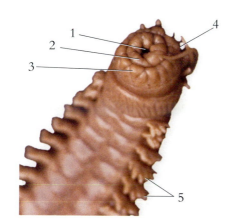

Figure 8.2 View of the head of the sandworm, *Neanthes*.

1. Mouth   4. Prostomium
2. Everted pharynx   5. Parapodia
3. Peristomium

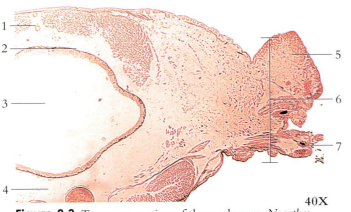

Figure 8.3 Transverse section of the sandworm, *Neanthes*.
1. Dorsal blood vessel
2. Intestine
3. Lumen of intestine
4. Coelom
5. Notopodium
6. Parapodium
7. Neuropodium

40X

Figure 8.4 Parapodium of the sandworm, *Neanthes*.
1. Dorsal cirrus
2. Notopodium
3. Striae
4. Neuropodium

100X

Class Oligochaeta

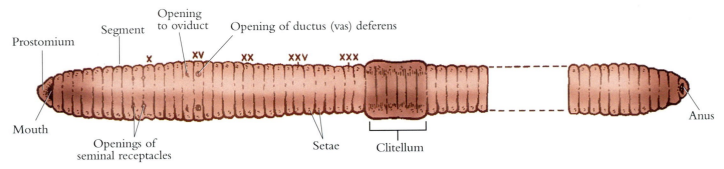

Figure 8.5 Ventral view diagram of the earthworm, *Lumbricus*.

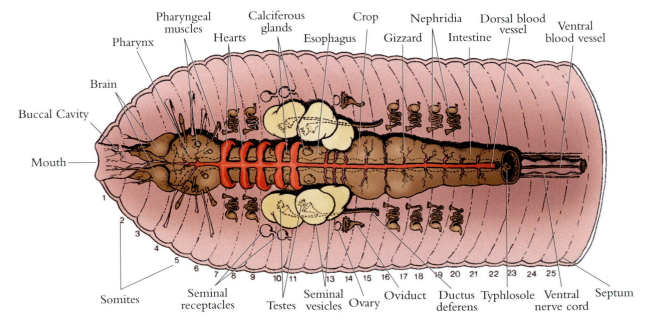

Figure 8.6 Diagram of the anterior end of the earthworm, *Lumbricus*.

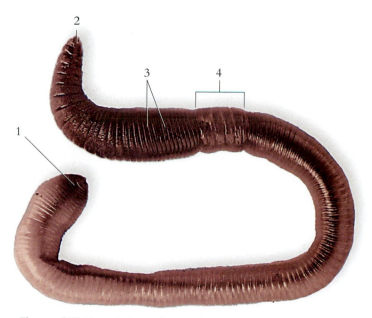

Figure 8.8 Anterior end of an earthworm, *Lumbricus*.

1. Prostomium   4. Opening of ductus (vas) deferens
2. Mouth   5. Segment 10
3. Setae

Figure 8.7 Dorsal view of an earthworm, *Lumbricus*.

1. Pygidium   3. Segments, or metameres
2. Prostomium   4. Clitellum

Figure 8.9 Earthworm cocoon (each line represents 1 mm).

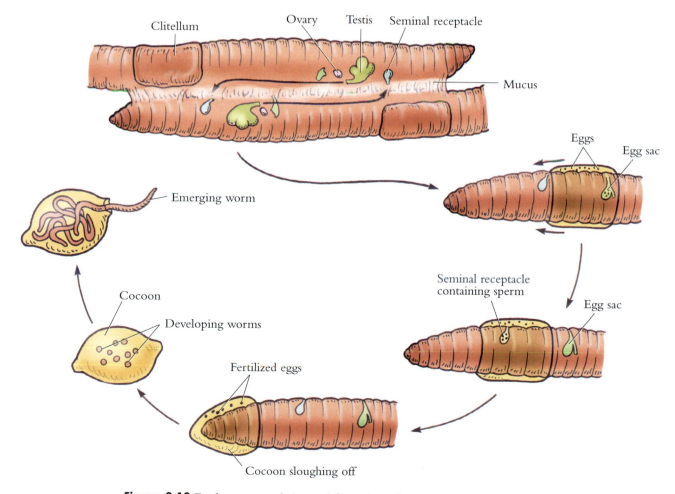

Figure 8.10 Earthworm copulation and formation of egg cocoon.

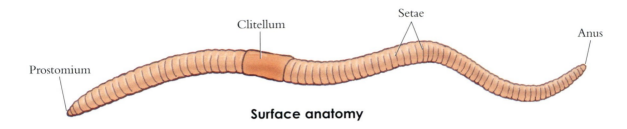

Surface anatomy

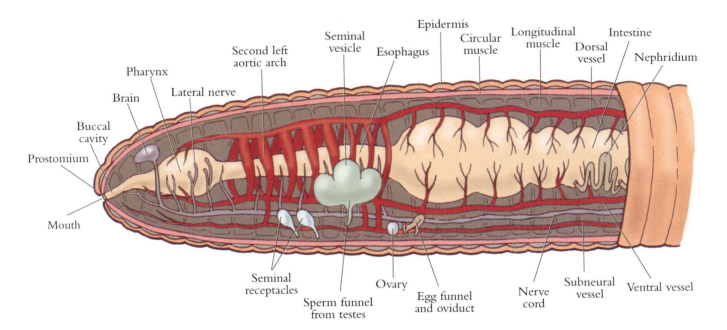

Longitudinal section

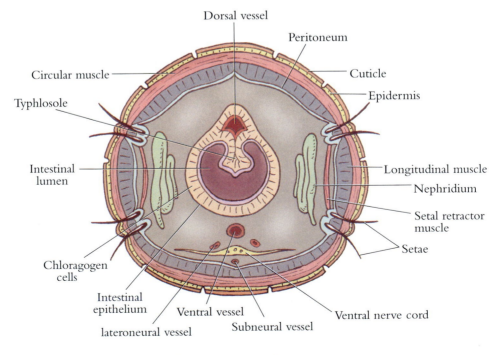

Transverse section

Figure 8.11 Anatomy of an earthworm.

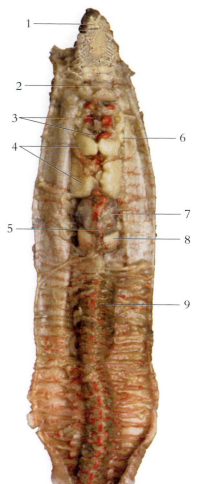

Figure 8.12 Internal anatomy of the anterior end of an earthworm, *Lumbricus*.
  1. Brain
  2. Pharynx
  3. Hearts
  4. Seminal vesicles
  5. Dorsal blood vessel
  6. Seminal recepticles
  7. Crop
  8. Gizzard
  9. Intestine

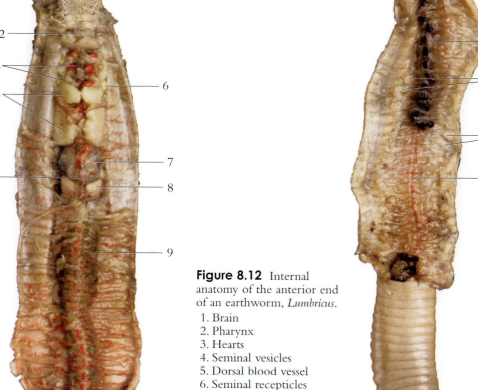

Figure 8.13 Internal anatomy of the posterior end of an earthworm with part of the intestine removed.
  1. Intestine
  2. Septae
  3. Nephridia
  4. Ventral blood vessel

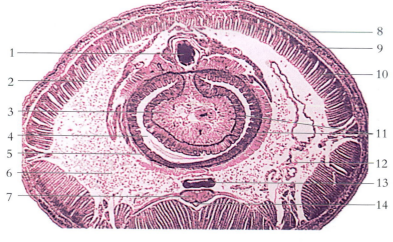

Figure 8.14 Transverse section of an earthworm posterior to the clitellum.
  1. Dorsal blood vessel
  2. Peritoneum
  3. Typhlosole
  4. Lumen of intestine
  5. Intestine
  6. Coelom
  7. Ventral nerve cord
  8. Epidermis
  9. Circular muscles
  10. Longitudinal muscles
  11. Intestinal epithelium
  12. Nephridium
  13. Ventral blood vessel
  14. Subneural blood vessel

Class Hirudinea

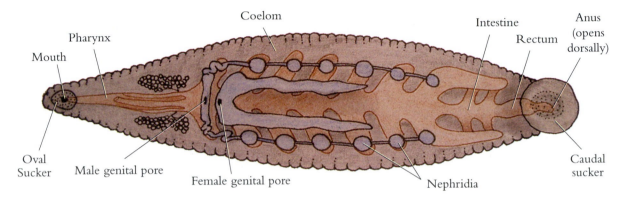

Figure 8.15 Diagram of a leech.

Figure 8.16 Leeches are more specialized than other annelids. They have lost their setae and developed suckers for attachment while sucking blood.

An estimated 80,000 species of roundworms are contained within the phylum *Nematoda*. Nematodes include a wide variety of small, elongated animals including free-living forms in water and soil, and parasitic forms such as hookworms, pinworms, *Ascaris*, and *Trichinella*.

Most of the nematodes are microscopic, bilaterally symmetrical, cylindrical and unsegmented, wormlike animals. The body of a nematode is enclosed in a tough cuticle that is shed periodically as the animal grows. Contraction of the longitudinal skeletal muscles attached to the cuticle causes a whiplike body movement. Nematodes lack circular muscle and have an incomplete mesodermal layer. The body cavity is called a *pseudocoelom* because it lacks an epithelial lining. The tubular digestive tract extends the length of the body from the mouth to the anus. The sexes are usually separate, and the female is larger than the male.

It is estimated that a spadeful of soil contains thousands of nematodes. They are the primary consumers of organic material and are also extremely important in subterranean and aquatic food chains. Other nematodes are parasites of plants and animals. Among the human nematode parasites are the hookworms, the intestinal roundworm *Ascaris*, pinworms, trichina worms (*Trichinella*), and filarial worms.

Nematodes, along with six other phyla of animals, have a pseudocoelom, and are sometimes collectively referred to as *pseudocoelomates* (see Figure 9.1). Because the pseudocoelom is fluid–filled or contains a gelatinous substance, it permits a greater freedom of movement as compared to the solid body structure of acoelomates. The pseudocoelom provides a space for development and specialization of digestive, reproductive, and excretory systems.

Although the pseudocoelomates are frequently discussed together in textbooks, they are polyphyletic (not derived from a common ancestor). The seven pseudocoelomate phyla are Nematoda, Rotifera, Gastrotricha, Kinorhyncha, Nematomorpha, Acanthocephala, and Entoprocta. Of these, the parasitic members within the phylum Nematoda are probably the most important to humans.

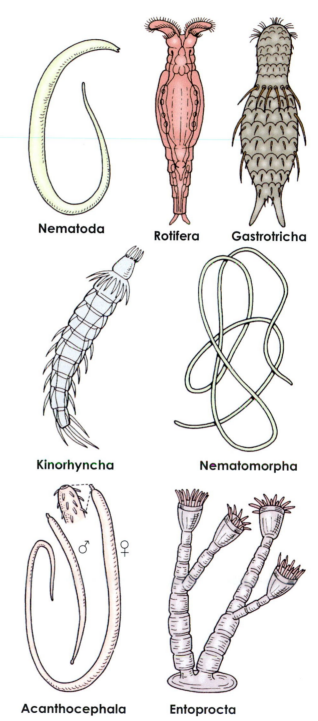

Nematoda **Rotifera** **Gastrotricha**

Kinorhyncha **Nematomorpha**

Acanthocephala **Entoprocta**

Figure 9.1 Representative pseudocoelomate phyla.

Phylum Nematoda

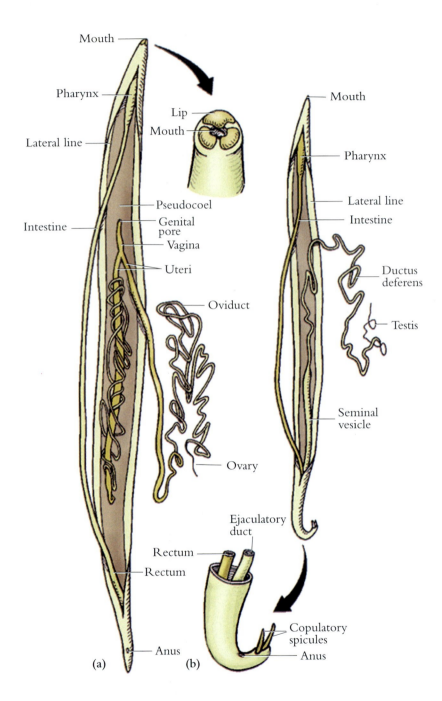

Mouth

Pharynx

Lateral line

Lip
Mouth

Intestine

Pseudocoel
Genital pore
Vagina
Uteri

Oviduct

Ovary

Mouth

Pharynx

Lateral line
Intestine

Ductus deferens

Testis

Seminal vesicle

Ejaculatory duct

Rectum
Rectum

Copulatory spicules
Anus

Anus

(a)   (b)

Figure 9.2 Diagrams of the internal anatomy of (a) a female and (b) a male *Ascaris*.

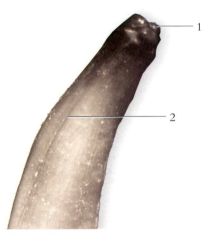

Figure 9.3 Head end of a male *Ascaris*.
1. Lip
2. Lateral line

(a)   (b)

Figure 9.4 Posterior end of (a) a female and (b) a male *Ascaris*.
1. Ejaculatory duct

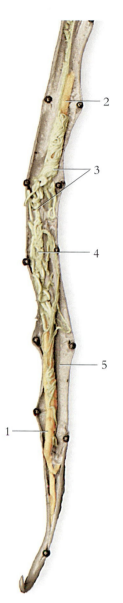

Figure 9.5 Internal anatomy of a male *Ascaris*.

1. Seminal vesicle
2. Intestine
3. Ductus deferens
4. Testes
5. Lateral line

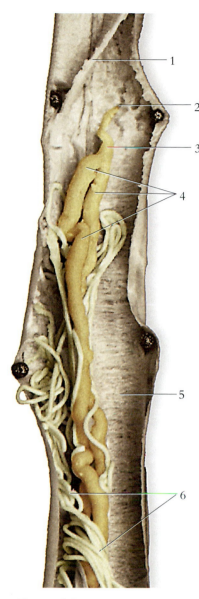

Figure 9.6 Internal anatomy of a female *Ascaris*.

1. Intestine
2. Genital pore
3. Vagina
4. Uteri (Y-shaped)
5. Lateral line
6. Oviducts

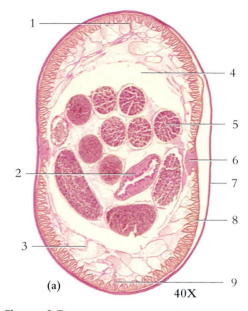

(a)   40X

Figure 9.7 Transverse sections of a male *Ascaris*.

1. Dorsal nerve cord	6. Lateral line
2. Intestine	7. Cuticle
3. Longitudinal muscle cell body	8. Contractile sheath of muscle cell
4. Pseudocoel	9. Ventral nerve cord
5. Testis	

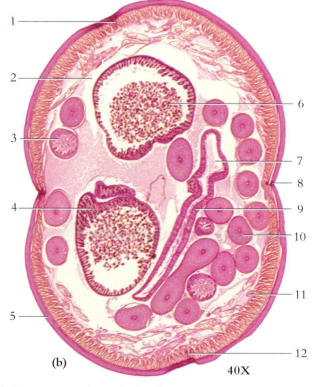

(b)   40X

Figure 9.8 Transverse sections of a female *Ascaris*.

1. Dorsal nerve cord	5. Cuticle	9. Intestine
2. Pseudocoel	6. Eggs	10. Ovary
3. Oviduct	7. Lumen of intestine	11. Longitudinal muscles
4. Uterus	8. Lateral line	12. Ventral nerve cord

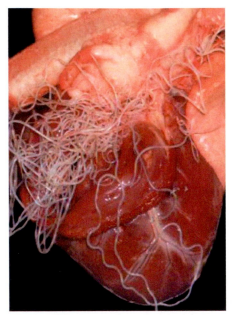

Figure 9.9 Dog heart infested with heartworm, *Dirofilaria immitis.*

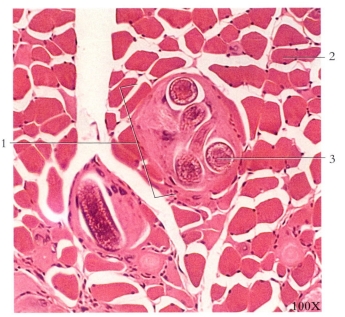

100X

Figure 9.10 Photomicrograph of *Trichinella spiralis* encysted in muscle.
1. Cyst   3. Larva
2. Muscle

Phylum Rotifera

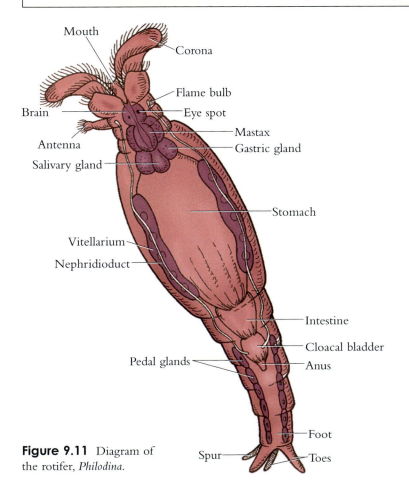

Mouth
Corona
Flame bulb
Brain
Eye spot
Antenna
Mastax
Salivary gland
Gastric gland
Stomach
Vitellarium
Nephridioduct
Intestine
Cloacal bladder
Pedal glands
Anus
Foot
Spur
Toes

Figure 9.11 Diagram of the rotifer, *Philodina.*

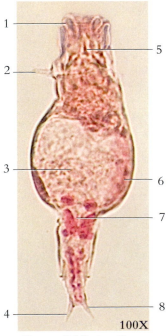

100X

Figure 9.12 Rotifer.
1. Corona   5. Mastax
2. Antenna   6. Vitellarium
3. Stomach   7. Intestine
4. Spur   8. Toe

Containing over 900,000 species, Arthropoda is the largest phylum within the kingdom Animalia. Arthropods are also the most biologically successful of all animals. There are more species of them, they live in a greater variety of habitats, and they eat a greater variety and amount of food than the members of any other phylum. Included within this phylum are diverse organisms such as horseshoe crabs, spiders, ticks, scorpions, lobsters, crabs, shrimp, insects, centipedes, and millipedes.

Arthropods have a *segmented body*, paired and highly specialized *jointed appendages, a chitinous exoskeleton* that is periodically shed as the animal grows, and an *open circulatory system* in which the blood that is pumped by the dorsally positioned heart flows through a cavity called a *hemocoel*. The sense organs are well developed in arthropods and most have highly specialized *compound eyes*. As compared to other invertebrates, arthropods have complex, innate (unlearned) behavior patterns.

The tremendous success of arthropods is due to the structural and physiological aspects of their body organization, and their reproductive potential. Most female arthropods, especially the insects, produce thousands of eggs during their life. The eggs generally hatch when food is abundant, and the young quickly develop through gradual metamorphosis or complete metamorphosis.

The economical importance of arthropods is immeasurable. Many insects are considered pests in that they feed upon human crops. Some arthropods, such as lobsters, crabs, and shrimp, are an important source of human food. Bees pollinate flowers and produce honey for human consumption. Most spiders are considered beneficial because they feed on noxious insects. Ticks and many insects, especially flies and mosquitoes, are of medical concern because they are vectors (carriers) of pathogenic microorganisms.

Table 10.1 Representatives of the Phylum Arthropoda

Classes and Representative Kinds	Characteristics
Merostomata — horseshoecrab	Cephalothorax and abdomen; specialized front appendages into chelicerae; lack antennae and mandibles
Arachnida — spiders, mites, ticks, and scorpions	Cephalothorax and abdomen; four pairs of legs; book lungs or trchea; lack antennae and mandibles
Malacostraca — lobsters, crabs, and shrimp	Cephalothorax and abdomen; two pairs of antennae; pair of mandibles and two pairs of maxillae; biramous appendages; gills
Maxillopoda — copepods and barnacles	Cephalothorax and abdomen; freshwater and marine; up to six pairs of appendages
Insecta — beetles, butterflies, and ants	Head, thorax, and abdomen; three pairs of legs; well-developed mouth parts; usually two pair of wings; trachea
Chilopoda — centipedes	Head with segmented trunk; one pair of legs per segment; trachea; one pair of antennae
Diplopoda — millipedes	Head with segmented trunk; usually two pair of legs per segment; trachea

Figure 10.1 Trilobite, *Modicia typicalis*, is an extinct arthropod from the Cambrian and Ordovician periods.

Class Merostomata

(a)

(b)

Figure 10.2 (a) Dorsal view and (b) a ventral view of the horseshoe crab, *Limulus*. This animal is commonly found in shallow waters along the Atlantic coast from Canada to Mexico.

1. Simple eye	6. Abdomen (opisthosoma)	11. Book gills	16. Anus
2. Compound eye	7. Telson	12. Pedipalp	17. Telson
3. Abdominal spines	8. Chelicerae	13. Mouth	
4. Anterior spine	9. Gnathobase	14. Chilaria	
5. Cephalothorax (prosoma)	10. Chelate legs	15. Genital operculum	

Class Arachnida

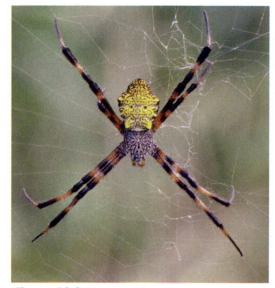

Figure 10.3 Beach spider , *Argiope appensa*, is an introduced species to Hawaii.

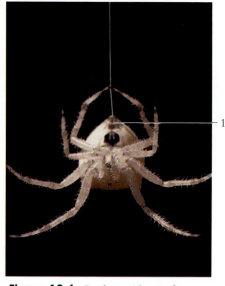

Figure 10.4 Garden spider in the process of spinning a web.
1. Spinnerets

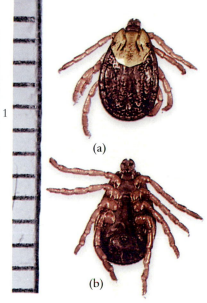

(a)

(b)

Figure 10.5 Ticks, within the family Ixodidae, are specialized parasitic arthropods. (a) A dorsal view and (b) a ventral view.

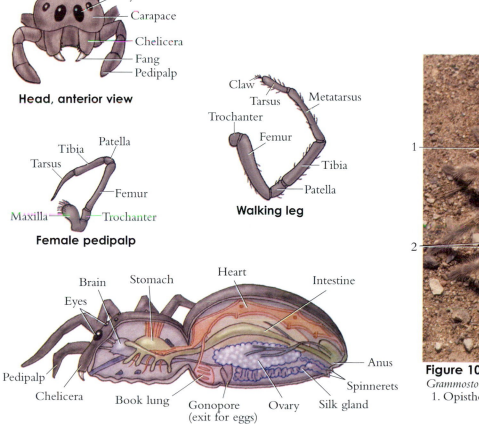

Eyes
Carapace
Chelicera
Fang
Pedipalp

Head, anterior view

Claw
Tarsus
Metatarsus
Trochanter
Femur
Tibia
Patella

Walking leg

Tibia
Patella
Tarsus
Femur
Maxilla
Trochanter

Female pedipalp

Brain
Stomach
Heart
Intestine
Eyes
Anus
Spinnerets
Pedipalp
Chelicera
Book lung
Gonopore (exit for eggs)
Ovary
Silk gland

Figure 10.6 Diagram of the anatomy of a spider.

Figure 10.7 Rose-haired tarantula, *Grammostola cola*.
1. Opisthosoma   2. Pedipalp   3. Prosoma

Figure 10.8 Scorpion, *Pandinus*. Scorpions are most commonly found in tropical and subtropical regions, but there are also several species found in temperate zones.

1. Pedipalp
2. Cephalothorax
3. Walking legs
4. Preabdomen
5. Stinging apparatus
6. Postabdomen (tail)

Figure 10.9 Ticks attached and feeding on a savannah monitor, a large African lizard.
1. Ticks   2. Scales of monitor

Class Malacostraca

Figure 10.10 Peppermint shrimp, *Lysmata wurdemanni*.

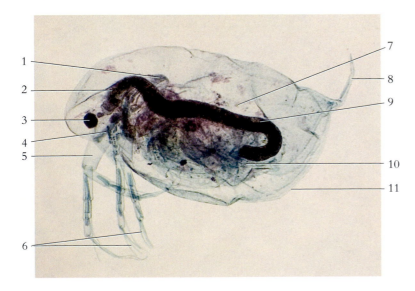

Figure 10.11 Ghost crab, *Ocypode ceratophthalmus*.

Figure 10.12 Water flea, *Daphnia*, is a common microscopic crustacean.

1. Heart
2. Midgut
3. Compound eye
4. Mouth
5. 2nd antenna
6. Setae
7. Brood chamber
8. Abdominal seta
9. Hindgut
10. Anus
11. Carapace

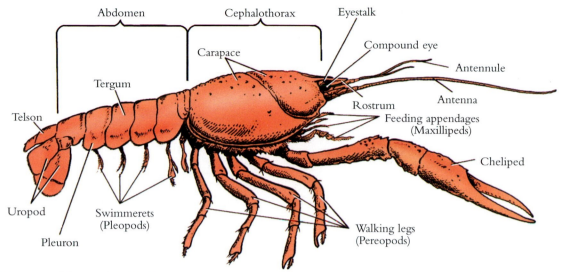

Figure 10.13 Diagram of the crayfish, *Cambarus*.

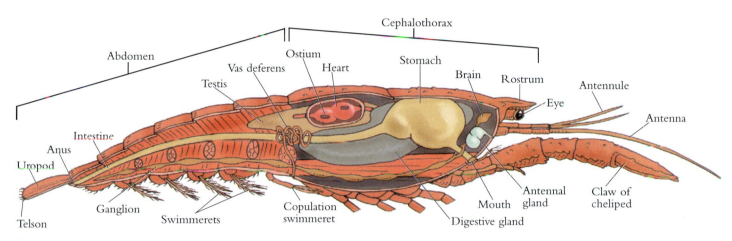

Figure 10.14 Anatomy of a crayfish. A sagittal section of an adult male.

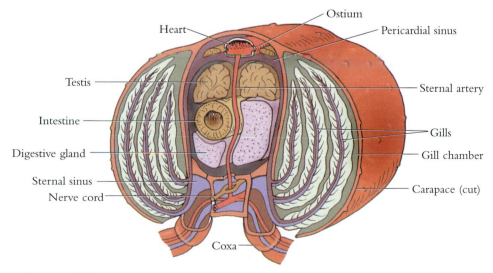

Figure 10.15 Anatomy of a crayfish. A transverse section of an adult male.

Figure 10.17 Lateral view of the crayfish.

1. Carapace
2. Abdomen
3. Uropod
4. Swimmeret (pleopod)
5. Rostrum
6. Eye
7. Maxilliped
8. Cheliped
9. Walking legs

Figure 10.16 Dorsal view of the crayfish.

1. Cheliped
2. Walking legs
3. Carapace
4. Abdomen
5. Telson
6. Uropod
7. Antenna
8. Antennule
9. Rostrum
10. Eye
11. Cephalothorax
12. Tergum

Figure 10.18 Ventral view of the oral region of the crayfish.

1. Third maxilliped
2. First maxilliped
3. Second maxilla
4. Green gland duct
5. Mandible

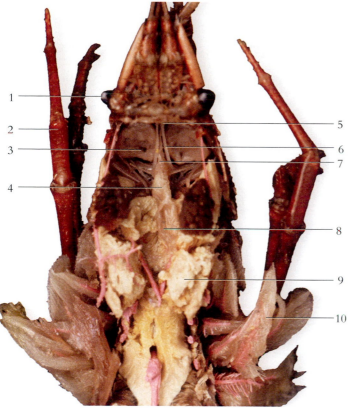

Figure 10.19 Dorsal view of the oral region of the crayfish.

1. Eye
2. Walking leg
3. Green gland
4. Cardiac chamber of stomach
5. Brain
6. Circumesophageal connection (of ventral nerve cord)
7. Esophagus
8. Region of gastric mill
9. Digestive gland
10. Gill

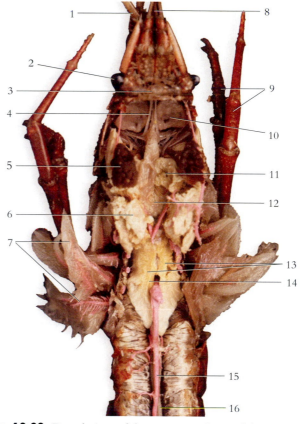

Figure 10.20 Dorsal view of the anatomy of a crayfish.

1. Antenna
2. Compound eye
3. Brain
4. Circumesophageal connection (of ventral nerve cord)
5. Mandibular muscle
6. Digestive gland
7. Gills
8. Antennules
9. Walking legs
10. Green gland
11. Esophagus
12. Pyloric stomach
13. Testis
14. Ductus deferens
15. Aorta
16. Intestine

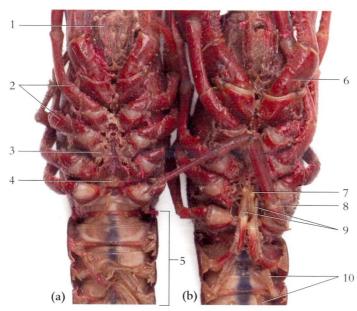

(a)   (b)

Figure 10.21 Ventral views of (a) a female and (b) a male crayfish. The first pair of swimmerets are greatly enlarged in the male for the depositing of sperm in the female's seminal receptacle.

1. Third maxilliped
2. Walking legs
3. Disc covering oviduct
4. Seminal receptacle
5. Abdomen
6. Base of cheliped
7. Sperm ducts (genital pores)
8. Base of last walking leg
9. Copulatory swimmerets (pleopods)
10. Swimmerets (pleopods)

Class Insecta

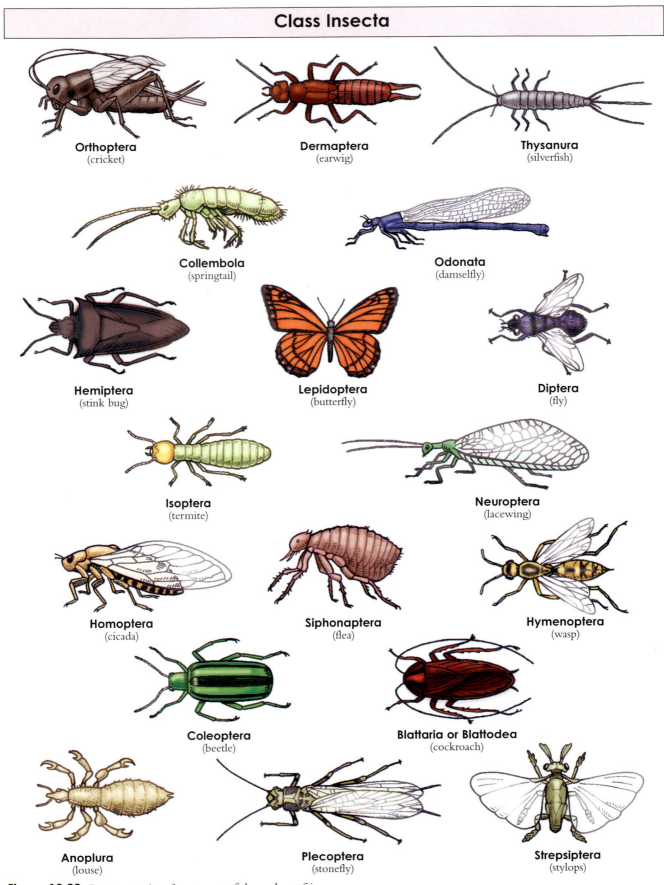

Orthoptera
(cricket)

Dermaptera
(earwig)

Thysanura
(silverfish)

Collembola
(springtail)

Odonata
(damselfly)

Hemiptera
(stink bug)

Lepidoptera
(butterfly)

Diptera
(fly)

Isoptera
(termite)

Neuroptera
(lacewing)

Homoptera
(cicada)

Siphonaptera
(flea)

Hymenoptera
(wasp)

Coleoptera
(beetle)

Blattaria or Blattodea
(cockroach)

Anoplura
(louse)

Plecoptera
(stonefly)

Strepsiptera
(stylops)

Figure 10.22 Representatives from some of the orders of insects.

Figure 10.23 Field cricket, *Gryllus firmus*.

Figure 10.24 Dragonfly, *Anax*.

Figure 10.25 Seven-spot ladybird beetle, *Coccinella septempunctata*.

Figure 10.26 Paper wasp, *Polistes*.

Figure 10.27 Praying mantis, *Stagmomantis*.

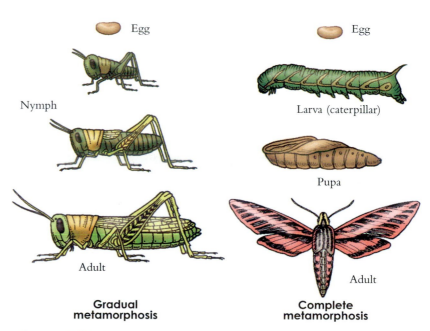

Figure 10.28 Insect development. In gradual (incomplete) metamorphosis the young resemble the adults but they are smaller and have different body proportions. In complete metamorphosis, the larvae look different than the adult and generally have different food requirements.

(a)

(b)

(c)

Figure 10.30 Developmental stages of the common honeybee include (a) larval stage, (b) pupa, and (c) adult.

Figure 10.29 Common grey cricket molting. All arthropods must periodically shed their exoskeleton in order to grow. This process is called molting, or ecdysis.

(a)

(b)

Figure 10.31
Hind legs of a honeybee worker, outer surface (a) and inner surface (b).
1. Coxa
2. Trochanter
3. Pollen basket
4. Pollen packer
5. Pecten
6. Femur
7. Tibia
8. Metatarsus
9. Pollen comb
10. Tarsus

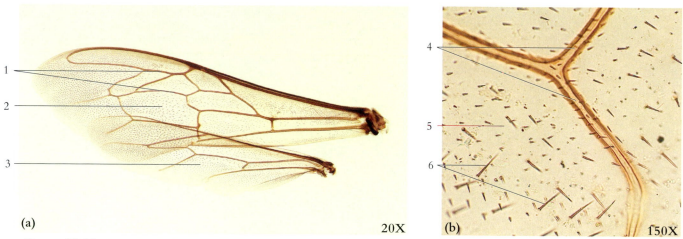

(a)   20X   (b)   150X

Figure 10.32 Wings of the honeybee, *Apis mellifera*. (a) Whole mount and (b) a close up.

1. Cross veins   4. Cross veins
2. Forewing   5. Transparent wing film
3. Hindwing   6. Hairs

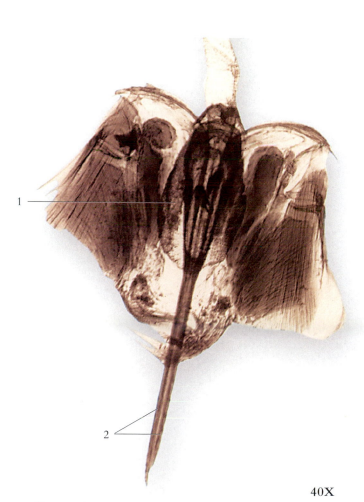

40X

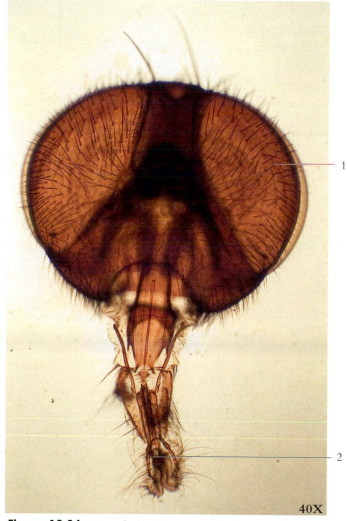

40X

Figure 10.33 Honeybee stinger. The two darts contain barbs on the tips which point upward, making it difficult to remove a stinger from a wound.

1. Sheath   2. Darts

Figure 10.34 Housefly head, an example of a sponging type of mouthpart in insects. Notice the large lobes at the apex of the labium, which function in lapping up liquids.

1. Compound eye   2. Labium

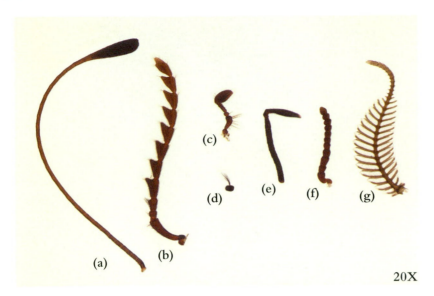

Figure 10.35 Common insect antennae. (a) Clavate – butterflies, (b) serrate – click beetles, (c) lamellate – scarab beetles, (d) aristate – house flies, (e) geniculate – weevils (f) moniliform – termites, and (g) plumose – moths.

20X

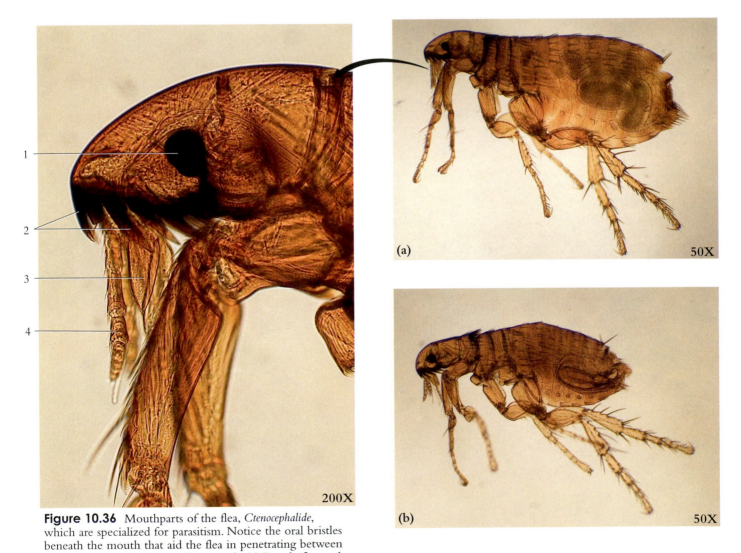

Figure 10.36 Mouthparts of the flea, *Ctenocephalide*, which are specialized for parasitism. Notice the oral bristles beneath the mouth that aid the flea in penetrating between hairs to feed on the blood of mammals. (a) Female flea and (b) male flea.

1. Eye
2. Oral bristles
3. Maxilla
4. Maxillary palp

(a)   50X

(b)   50X

200X

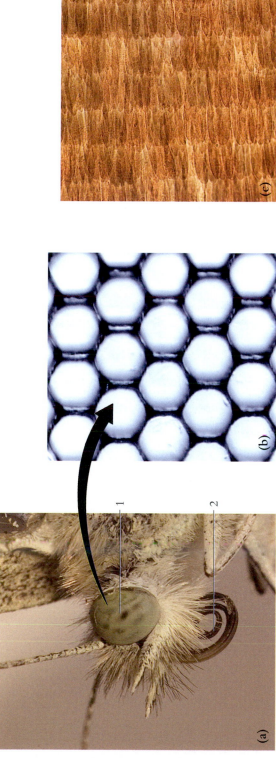

Figure 10.37 (a) Lateral view of the head of a butterfly. The most obvious structures on the head of a butterfly are compound eyes and the curled tongue for siphoning nectar from flowers. (b) A magnified view of the compound eye, and (c) a magnified view of the wing scales.

1. Compound eye   2. Tongue

Figure 10.39 Preserved specimen of a grasshopper, order Orthoptera.

1. Mesothorax	7. Maxilla	13. Mesothorax
2. Vertex	8. Antenna	14. Spiracle
3. Compound eye	9. Claw	15. Abdomen
4. Vertex	10. Wing	16. Trochanter
5. Gena	11. Femur	17. Tarsus
6. Frons	12. Tibia	

Figure 10.38 Surface anatomy of a grasshopper, order Orthoptera.

1. Antenna	7. Labrum	13. Tergum
2. Compound eye	8. Labial palp	14. Spiracle
3. Ocelli	9. Femur	15. Cercus
4. Prothorax	10. Metathorax	16. Ovipositor
5. Mesothorax	11. Tibia	17. Tarsus
6. Tympanum	12. Forewing	18. Sternum

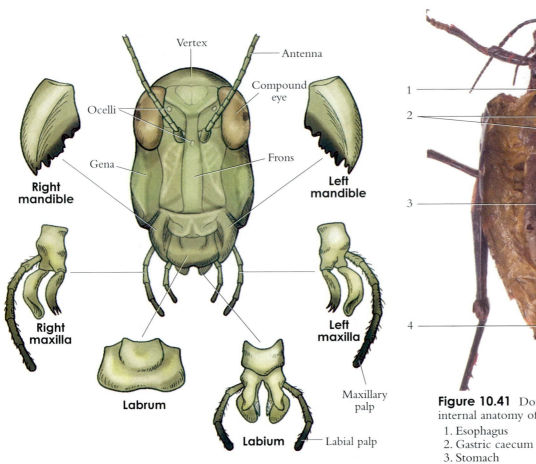

Figure 10.40 Diagram of the head and mouthparts of a grasshopper.

Labels in figure: Vertex, Antenna, Compound eye, Ocelli, Gena, Frons, Right mandible, Left mandible, Right maxilla, Left maxilla, Labrum, Labium, Maxillary palp, Labial palp

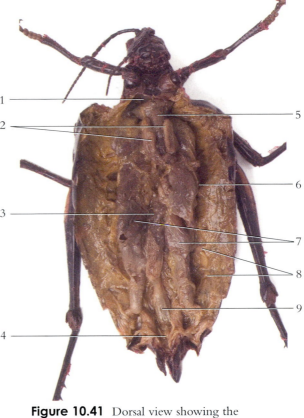

Figure 10.41 Dorsal view showing the internal anatomy of a grasshopper.

1. Esophagus
2. Gastric caecum
3. Stomach
4. Rectum
5. Crop
6. Malpighian tubule
7. Ovarys
8. Trachea
9. Intestine

Class Chilopoda

Figure 10.42 Garden centipede, *Lithobius*.

Class Diplopoda

Figure 10.43 Giant african millipede, *Scaphiostreptus*.

An estimated 6,000 species of echinoderms are contained within the phylum *Echinodermata*. Echinoderms are bottom-dwelling marine animals that include such forms as sea stars (star fish), sea urchins, sand dollars, sea cucumbers, and crinoids.

Echinoderms have a well-developed coelom, most have a complete digestive system, a rudimentary circulatory system, a simple nervous system consisting of nerve rings surrounding the mouth with radiating nerves, and various types of respiratory structures (e.g., dermal gills in sea stars; respiratory trees in sea cucumbers). In some, the endoskeleton is of small calcareous plates that bear outwardly projecting spines. Echinoderms have a water-vascular system composed of a network of canals through which seawater circulates. Branches of this system lead to numerous tube feet, which extend when filled with water. The tube feet function in locomotion, obtaining food, and in some forms, gas exchange. Most echinoderms have regenerative capacities, and an entire sea star can regenerate from a single severed arm.

Echinoderms are extremely important in the marine food chain. Sea urchins are food for many organisms, ranging from sea stars to sea otters. Many people in Asia consume large quantities of sea cucumbers and the eggs of sea urchins. Because sea stars commonly feed on clams, oysters, and other mollusks, they are considered a threat to the shellfish industry.

Echinoderms are also important laboratory animals used in developmental biology experiments. Their gametes are easily harvested and are sufficiently large that they can be readily studied.

Table 11.1 Representatives of the Phylum Echinodermata

Classes and Representative Kinds	Characteristics
Asteroidea — sea stars (star fish)	Pentaradial symmetry; appendages arranged around a central disk containing the mouth; tube feet with suckers
Echinoidea — sea urchins, and sand dollars	Disk-shaped with no arms; compact skeleton; movable spines; tube feet with suckers
Ophiuroidea — brittle stars	Pentaradial symmetry; appendages sharply marked off from central disk; tube feet without suckers
Holothuroidea — sea cucumbers	Cucumber-shaped with no arms; spines absent; tube feet with tentacles and suckers
Crinoidea — sea lilies and feather stars	Sessile during much of life cycle; calyx supported by elongated stalk in some

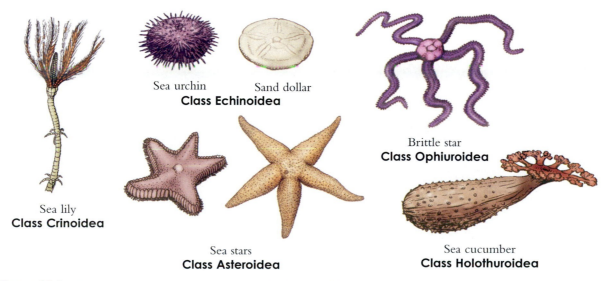

Sea urchin   Sand dollar
Class Echinoidea

Brittle star
Class Ophiuroidea

Sea lily
Class Crinoidea

Sea stars
Class Asteroidea

Sea cucumber
Class Holothuroidea

Figure 11.1 Diagram of specimens representing each class of echinoderms.

Figure 11.2   Green sea urchin, *Strongylocentrotus droebachiensis*.

Figure 11.3   Red slate sea urchin, *Heterocentrotus mammillatus*.

Figure 11.4   Sea stars (star fish), *Asterias*, grouped in a tide pool in southern California.

Figure 11.5   Sea star (star fish), *Asterias* .

Figure 11.6   Spiny brittle stars, *Ophiothix spiculata*, on a sea wall.

Figure 11.7   California sea cucumber, *Parastichopus californicus*.

Class Asteroidea

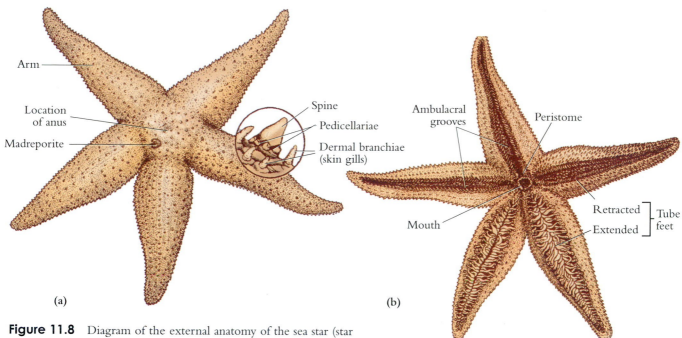

Figure 11.8 Diagram of the external anatomy of the sea star (star fish), *Asterias*. (a) An aboral (dorsal) view and (b) an oral (ventral) view.

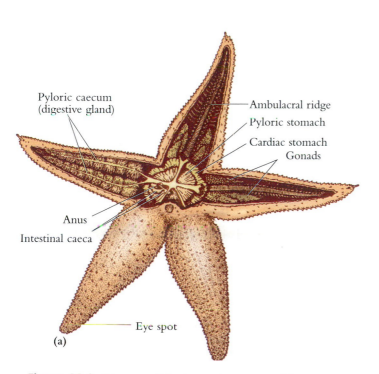

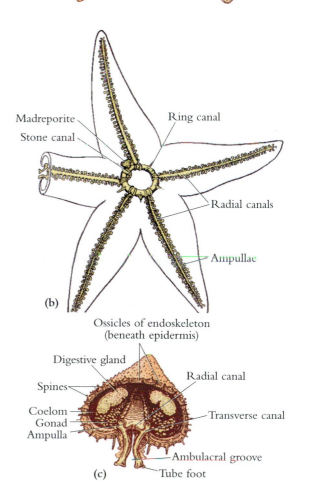

Figure 11.9 Diagram of the internal anatomy of the sea star from aboral side. (a) The digestive and reproductive organs, (b) the water vascular system, and (c) a transverse section through an arm.

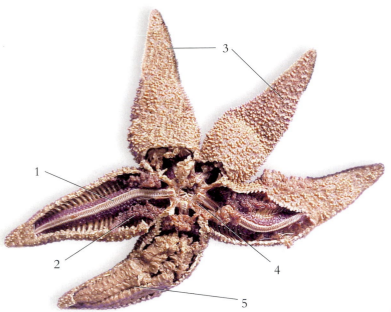

Figure 11.10 Aboral view of the internal anatomy of a sea star.
1. Ambulacral ridge
2. Gonad
3. Spines
4. Circular canal
5. Pyloric caecum (digestive gland)

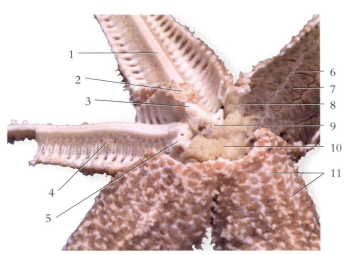

Figure 11.11 Magnified aboral view of the internal anatomy of a sea star.
1. Polian vesicle
2. Ampullae
3. Stone canal
4. Ambulacral ridge
5. Madreporite
6. Pyloric duct
7. Pyloric caecum (digestive gland)
8. Gonad
9. Anus
10. Pyloric stomach
11. Spines

Figure 11.12 Oral view of a sea star.
1. Tube feet
2. Peristome
3. Mouth
4. Ambulacral groove
5. Oral spines

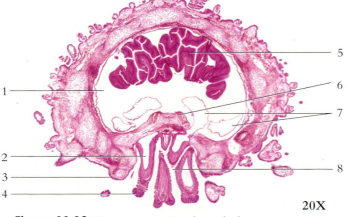

20X

Figure 11.13 Transverse section through the arm of a sea star.
1. Coelom
2. Tube foot
3. Epidermis
4. Sucker
5. Pyloric caecum
6. Radial canal
7. Gonad
8. Ambulacral groove

Class Echinoidea

Figure 11.14 Oral view of the sea urchin, *Arbacia*.
1. Mouth   3. Pedicellaria
2. Spines   4. Peristome

Figure 11.15 Aboral view of the sea urchin.
1. Ossicles   2. Madreporite

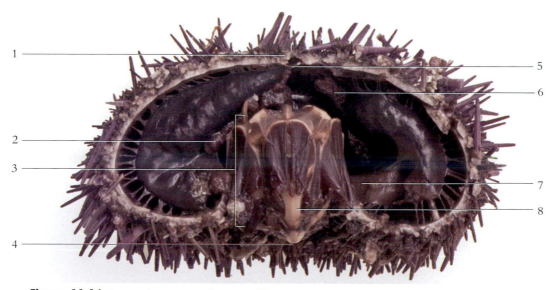

Figure 11.16 Internal anatomy of a sea urchin.
1. Madreporite   5. Anus
2. Intestine   6. Gonad
3. Aristotle's lantern 7. Stomach
4. Mouth   8. Calcareous tooth

Class Holothuroidea

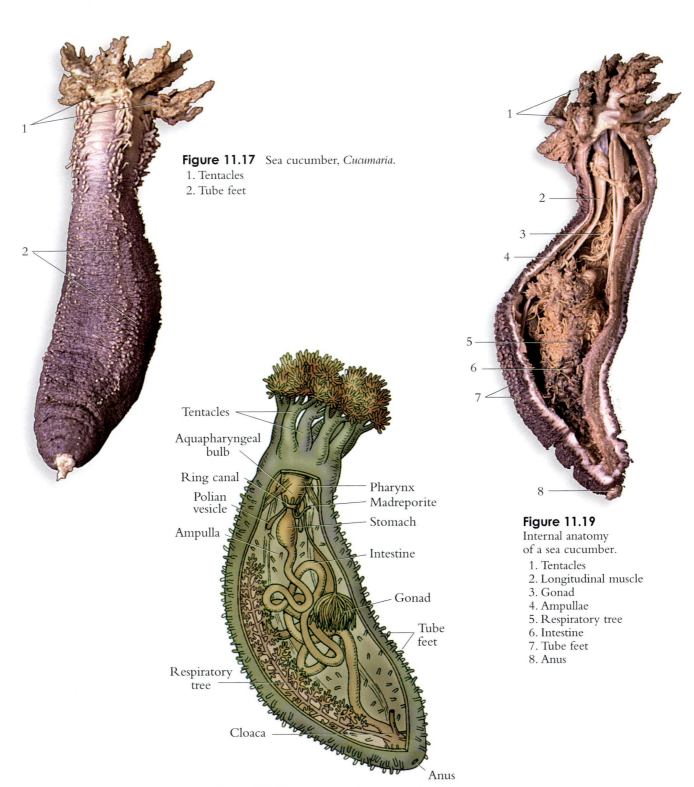

Figure 11.17 Sea cucumber, *Cucumaria*.
1. Tentacles
2. Tube feet

Tentacles
Aquapharyngeal bulb
Ring canal
Polian vesicle
Ampulla
Pharynx
Madreporite
Stomach
Intestine
Gonad
Tube feet
Respiratory tree
Cloaca
Anus

Figure 11.18 Diagram of the internal anatomy of a sea cucumber.

Figure 11.19
Internal anatomy of a sea cucumber.

1. Tentacles
2. Longitudinal muscle
3. Gonad
4. Ampullae
5. Respiratory tree
6. Intestine
7. Tube feet
8. Anus

Chapter 12
Chordata

An estimated 45,000 species of chordates are contained within the phylum *Chordata*. Chordates include a wide variety of animals that have, during some stage of development, a notochord, pharyngeal gill slits, dorsal hollow nerve cord, and a postanal tail. Included in this phylum are the tunicates, the lancelets, and the vertebrate animals.

Chordates are thought to have descended from echinoderm-like ancestors during the Precambrian period. Tunicates are small, primitive chordates that are filter-feeders of plankton within marine waters. Lancelets are small fish-like chordates that inhabit sandy bottoms of coastal waters. The four chordate characteristics are present and functional in adult lancelets. Vertebrates are chordates that have a well-developed brain and a cartilaginous or bony vertebral column surrounding a dorsal nerve cord. There are an estimated 43,000 species of living vertebrates within eight classes — *Agnatha* (jawless fishes), *Chondrichthyes* (cartilaginous fishes), *Osteichthyes* (bony fishes), *Amphibia* (amphibians), *Reptilia* (reptiles), *Aves* (birds), and *Mammalia* (mammals).

Chordates are a highly specialized and successful group of animals. Generally large in size, they occupy diverse habitats and have specialized niches and behaviors. All chordates are important in the biological food chains and ecological communities.

Table 12.1 Representatives of the Phylum Chordata

Subphyla and Representative Kinds	Characteristics
Urochordata — tunicates	Marine, larvae are free-swimming and have notochord, gill slits, and dorsal hollow nerve cord; most adults are sessile (attached), filter-feeders, saclike animals
Cephalochordata — lancelets (*Amphioxus*)	Marine, segmented, elongated body with notochord extending the length of the body; cilia surrounding the mouth for obtaining food
Vertebrata — agnathans (lampreys and hagfishes), fishes (cartilaginous and bony), amphibians, reptiles, birds, mammals	Aquatic and terrestrial forms; distinct head and trunk supported by a series of cartilaginous or bony vertebrae in the adult; closed circulatory system and ventral heart; well-developed brain and sensory organs

Table 12.2 Representatives of the Subhylum Vertebrata

Taxa and Representative Kinds	Characteristics
Superclass Agnatha	Eel-like and aquatic; sucking mouth (some parasitic); lack jaws and paired appendages
Class Myxini — hagfishes	Terminal mouth with buccal funnel absent; nasal sac connected to pharynx; 4 pairs of tentacles; 5 to 10 pairs pharyngeal pouches
Class Cephalaspidomorphi — lampreys	Suctorial mouth with rasping teeth; nasal sac not connected to buccal cavity; 7 pairs of pharyngeal pouches
Superclass Gnathostomata	Jawed vertebrates; most with paired appendages
Class Chondrichthyes — sharks, rays, and skates	Cartilaginous skeleton; placoid scales; most have spiracle; spiral valve in digestive tract
Class Osteichthyes — bony fishes	Gills covered by bony operculum; most have swim bladder
Class Amphibia — salamanders, frogs, and toads	Larvae have gills and adults have lungs; scaleless skin (except apoda); an incomplete double circulation; three–chambered heart
Class Reptilia — turtles, snakes, and lizards	Amniotic egg; epidermal scales; three- or four-chambered heart; lungs
Class Aves — birds	Homeothermous (warm-blooded); feathers; toothless; air sacs; four-chambered heart with right aortic arch
Class Mammalia — mammals	Homeothermous; hair; mammary glands; most have seven cervical vertebrae; muscular diaphragm; three auditory ossicles; four–chambered heart with left aortic arch

Subphylum Urochordata

Figure 12.1 Adult tunicate, *Ciona intestinalis*.

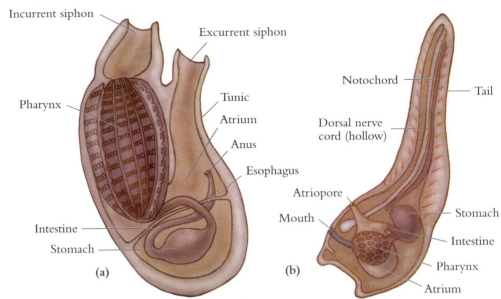

Figure 12.2 Diagram of a tunicate, (a) adult and (b) a larva.

Subphylum Cephalochordata

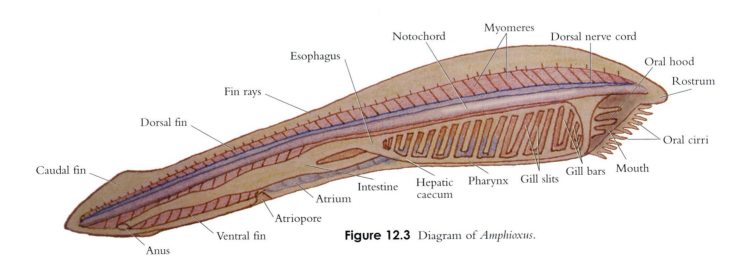

Figure 12.3 Diagram of *Amphioxus*.

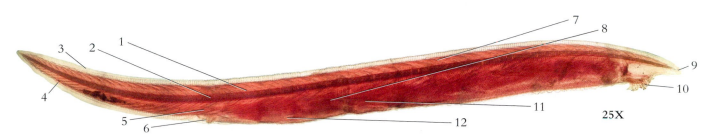

Figure 12.4 Whole mount of *Amphioxus*.

1. Dorsal nerve cord
2. Notochord
3. Caudal fin
4. Anus
5. Intestine
6. Atriopore
7. Myomeres
8. Esophagus
9. Rostrum
10. Oral cirri
11. Hepatic caecum
12. Atrium

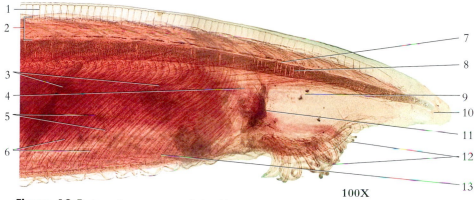

Figure 12.5 Anterior anatomy of *Amphioxus*.

1. Fin rays
2. Myomeres
3. Pigment spots
4. Velum
5. Gill slits
6. Gill bars
7. Dorsal nerve cord
8. Notochord
9. Rostrum
10. Oral hood
11. Wheel organ
12. Oral cirri
13. Pharynx

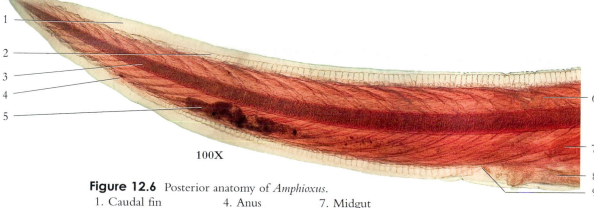

Figure 12.6 Posterior anatomy of *Amphioxus*.

1. Caudal fin
2. Fin rays
3. Notochord
4. Anus
5. Intestine
6. Myomeres
7. Midgut
8. Atrium
9. Atriopore

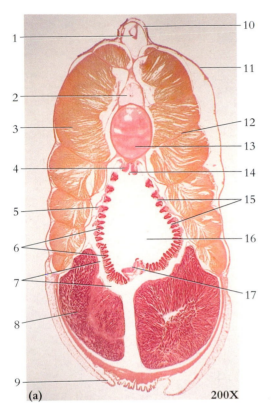

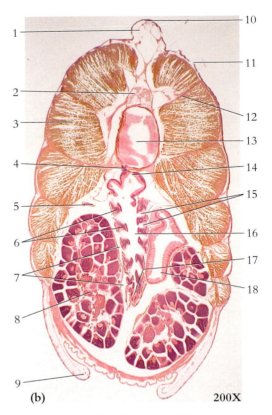

(a) 200X

(b) 200X

Figure 12.7 Transverse section through the pharyngeal region of (a) a male, and (b) a female *Amphioxus*.

1. Fin ray
2. Dorsal nerve cord
3. Myomere
4. Dorsal aorta
5. Nephridium
6. Gill bars
7. Atrium

8. Testis (male)
   Ovary (female)
9. Metapleural fold
10. Dorsal fin
11. Epidermis
12. Myoseptum
13. Notochord

14. Epibranchial groove
15. Gill slits
16. Pharynx
17. Endostyle (hypobranchial groove)
18. Hepatic caecum (liver)

Subphylum Vertebrata   Class Cephalaspidomorphi

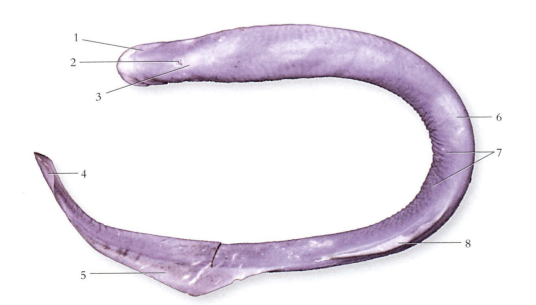

Figure 12.8 External anatomy of a marine lamprey, *Petromyzon marinus*.

1. Head
2. Nostril
3. Pineal body
4. Caudal fin
5. Posterior dorsal fin
6. Trunk
7. Myomeres
8. Anterior dorsal fin

Figure 12.9 Anterior anatomy of a marine lamprey.
1. Eye
2. Buccal funnel
3. External gill slits

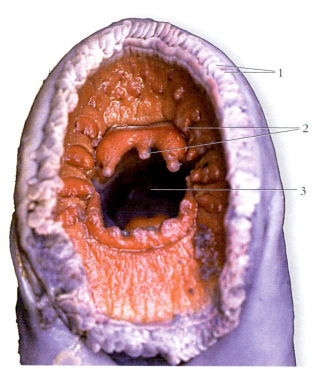

Figure 12.10 Oral region of a marine lamprey.
1. Buccal papillae
2. Horny teeth
3. Mouth

Figure 12.11 Cartilaginous skeleton of a marine lamprey.
1. Buccal cavity
2. Cranium
3. Lingual cartilage
4. Branchial basket
5. Notochord

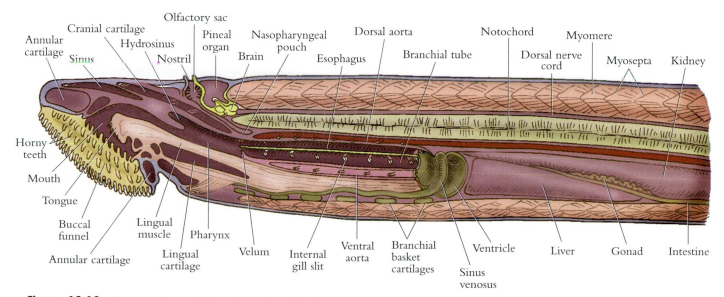

Figure 12.12 Diagram of a sagittal section of a marine lamprey.

Figure 12.13 Sagittal section through the anterior region of a lamprey.

1. Pineal organ	5. Mouth	9. Esophagus	13. Dorsal aorta
2. Nostril	6. Buccal muscle	10. Myomeres	14. Pericardial cavity
3. Brain	7. Lingual cartilage	11. Dorsal nerve cord	15. Sinus venosus
4. Annular cartilage	8. Pharynx	12. Notochord	16. Internal gill slits

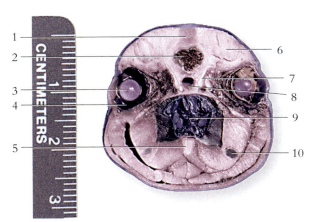

Figure 12.14 Transverse section through the eyes of a lamprey.

1. Pineal organ	7. Cranial cartilage
2. Brain	8. Nasopharyngeal pouch
3. Lens of eye	9. Pharynx
4. Retina of eye	10. Pharyngeal gland
5. Lingual cartilage	
6. Myomere	

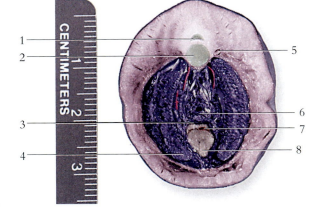

Figure 12.15 Transverse section through the branchial tube anterior to the fourth gill pouch of a lamprey. The ventral aorta is paired at this location.

1. Dorsal nerve cord	5. Anterior cardinal vein
2. Notochord	6. Esophagus
3. Branchial tube	7. Ventral aorta
4. Gill pouch	8. Lingual muscle

Subphylum Vertebrata   Class Chondrichthyes

Figure 12.16 Dorsal view of a skate, *Raja*.
1. Rostrum   3. Spiracle   5. Tail
2. Eyes   4. Pectoral fin 6. Pelvic fin

Figure 12.17 Detailed view of the head of a skate, *Raja*.
1. Eye
2. Spiracle
3. Rostrum

Figure 12.18
Lateral view of the leopard shark, *Triakis semifasciata*.
1. Spiracle
2. Lateral line
3. Dorsal fins
4. Caudal fin (heterocercal tail)
5. Eye
6. Gill slits
7. Pectoral fin
8. Pelvic fin
9. Anal fin

Subphylum Vertebrata   Class Osteichthyes

Eye   Lateral line   Dorsal fin   Caudal fin

Nostril

Mandible

Maxilla   Operculum   Pectoral fin   Pelvic fin   Anal fin

Figure 12.19 External structures of a grouper, *Mycteroperca bonaci*.

Figure 12.20 Lionfish, stonefish, and scorpionfish, within the family *Scorpaenidae*, have specialized spines in their dorsal fins for delivey of venom.

Figure 12.21 Tomato clownfish, *Amphiprion melanopus*, work as a pair to protect a group of eggs attached to rock surface.
  1. Egg mass

1

Subphylum Vertebrata   Class Amphibia

Figure 12.22 Eastern newt, *Notophthalmus viridescens,* lives in humid sites, often beneath debris along stream banks.

(a)

Figure 12.23 Surface anatomy and body regions of the leopard frog, *Rana pipiens.*

1. Ankle
2. Knee
3. Foot
4. Eyes
5. Nostril
6. Tympanic membrane
7. Brachium
8. Antebrachium
9. Digits

(b)

Figure 12.24 Adult Indonesian giant tree frog, *Litoria infrafrenata.* (a) The frog is crouched on a person's fingers. (b) The adhesive toe disks can be seen in a ventral view.

Figure 12.25 Poison-dart frog, *Dendrobates,* secretes a toxic poison from the skin. South American indians have used this poison to coat darts.

Subphylum Vertebrata   Class Reptilia

Figure 12.26 Photograph showing hatching California king snakes, *Lampropeltis gentulus*. Most snakes are oviparous, meaning they lay eggs such as these. Some snakes, including all American pit vipers, are ovoviviparous, giving birth to well-developed young.

Figure 12.27 Aldabra giant tortoise, *Dipsochelys dussumieri*.

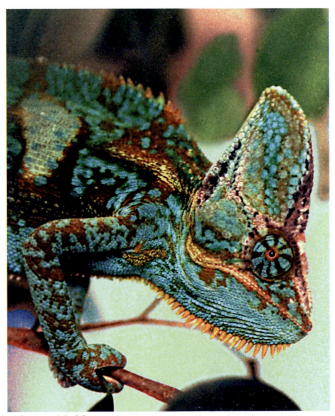

Figure 12.28 Veiled chameleon, *Chamaeleo calyptratus*. Chameleons are best known for their ability to change colors according to their surroundings.

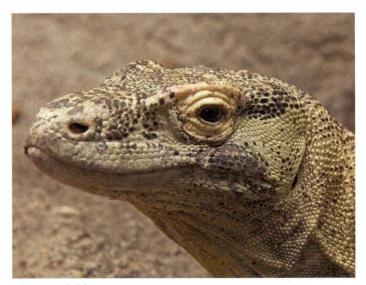

Figure 12.29 Komodo dragon, *Varanus komodoensis*, is the world's largest lizard species.

Figure 12.30 Garter snake, *Thamnophis sirtalis*.

Subphylum Vertebrata Class Aves

Figure 12.31 American kestrel, *Falco sparverius*, is common throughout North America. It hunts small rodents and insects.

Figure 12.32 Robin, *Turdus migratorius*, is a member of the trush family. It is the state bird of Wisconsin.

Figure 12.33 Magellanic penguins, *Spheniscus magellanicus*, are the most common of the 17 species of penguins.

Figure 12.34 Gila woodpecker, *Melanerpes uropygialis*, "anting". A behavior where the woodpecker allows ants to crawl through its feathers and eat any parasites that the bird might have.

Subphylum Vertebrata   Class Mammalia

Figure 12.35 Museum prepared mammal skins are important in mammalian taxonomy.

Figure 12.36 Mountain lion, *Felis concolor*, is also known as the cougar, puma, or panther. It is found throughout the western United States.

Figure 12.37 Malaysian fruit bat, *Pteropus hypomelanus*.

Figure 12.38 Capybara, *Hydrochaeris hydrochaeris*, is the world's largest rodent reaching a length of over 4 feet and weighing as much as 145 pounds.

Fishes are the vertebrate organisms most highly adapted for an aquatic environment. Many of the specialized structures of fishes enable excellent exploitation of their dense water medium in which they are confined. Gills enable extraction of oxygen from the water. Buoyancy is maintained through physiological adaptations that may or may not involve swim-bladders. Streamlined locomotion is achieved by body shapes, positions of fins, and placement of muscle masses. The sense organs of fishes are highly developed to respond to stimuli passing through a water medium.

Fishes include the organisms in three classes of vertebrates. Class *Agnatha* includes the jawless fishes (see Chapter 12), such as the lampreys and hagfishes. Class *Chondrichthyes* includes the cartilaginous fishes, such as the skates, rays, and sharks. The bony fishes include the ancient lobe-fin fishes (*Sarcopterygii*) and the more modern ray-fin fishes(*Actinopterygii*). The cartilaginous fishes and the bony fishes are described and depicted in this chapter.

Chondrichthyes (Cartilaginous Fishes)

Nearly 850 species of skates, rays, sharks, and chimaeras are contained within the class Chondrichthyes. The subclass *Elasmobranchii* contains the skates, rays, and sharks, which number 800 species. The subclass *Holocephali* contains the 25 species of chimaeras. Chimaeras are remnants of ancient elasmobranchs that were much more common during the Devonian period nearly 300 million years ago.

Some of the characteristics of Chondrichthyes include:

1. *Fusiform body* with a *heterocercal tail*. The body of most sharks is tapered at both ends — like the fuselage of an airplane. This streamlined shape is ideal for rapid swimming. The caudal vertebrae turn upward forming the heterocercal tail. Paired pectoral and pelvic fins are controlled by muscles attached to the girdles. Prominent dorsal fins provide stability while swimming.

2. Ventral-positioned mouth and a *spiracle* exiting from the oral cavity. Well–developed jaws surround the mouth and support many *homodont teeth* (similarily shaped teeth). Most Chondrichthyes are excellent predators.

3. *Placoid scales* cover the skin. Composed of enamel and dentin, the placoid scales provide excellent protection to these animals. The teeth of Chondrichthyes are modified placoid scales that are continuously replaced as they are shed.

4. *Cartilaginous endoskeleton*. Although derived from early Devonian ancestors that had bony endoskeletons, Chondrichthyes maintain a cartilaginous endoskeleton even as adult animals.

5. Five to seven pairs of *gills*. Except in chimaeras, the gills of Chondrichthyes are not covered by an operculum.

6. Two-chambered heart with several pairs of aortic arches and a dorsal and ventral aorta. High concentrations of urea and trimethylamine oxide are maintained within the blood that aid buoyancy. Chondrichthyes lack swim bladders and lungs.

7. *Spiral valve* in digestive system. The digestive system has distinct portions for specialized functions. The spiral valve delays the passage of food and increases the absorptive surface. Attached to the rectum is the *rectal gland* which secretes a fluid containing sodium chloride that assists the kidney in regulating the salt concentration of the blood.

8. Sensory portions of brain are well-developed. Chondrichthyes have a specialized *lateral line system* for detection of water vibrations. The lateral line system connects to the three pairs of semicircular canals. Olfaction is well-developed, but vision is considered poor.

9. Males have *claspers* and fertilization is internal, with *oviparous*, *ovoviviparous*, or *viviparous* development. Cartilaginous fishes have *mesonephros*–type kidneys.

Osteichthyes (Bony Fishes)

Class Osteichthyes includes the bony fishes of which there are approximately 24,600 species within the subclass *Sarcopterygii* (lobe-fin fishes) and subclass *Actinopterygii* (ray-fin fishes). There are few living species of lobe-fin fishes, but the group is highly significant because from ancient extinct members of these fishes descended the amphibians and indeed all tetrapod vertebrates. Lobe-fin fishes have bony elements supporting their paired pectoral and pelvic fins. Lungfishes within the group respire through gills as well as lungs. They also have nostrils that open into the mouth.

Some authorities present the classification of bony fishes into three subclasses rather than two. In this scheme, subclass *Crossopterytgii* is the designation of the lobe-fin fishes which contains one living order and one species, *Latimeria chalumnae*. Subclass *Dipneusti* contains two living orders of lungfishes and three genera: *Neoceratodus*, *Lepidosiren*, and *Protopterus*. All of the remaining bony fishes belong to the Subclass *Actinopterygii*, the ray-fin fishes.

Some of the characteristics of actinopterygians (ray–finned fishes) include:

1. Most have *fusiform body* with *homocercal tail* (uniform arrangement of supporting rays).

2. *Dermal scales* present within glandular skin. The three kinds of scales are *ganoid*, *cycloid*, and *ctenoid*. A few bony fishes lack scales. *Mucous glands* are abundant within the skin and secrete a protective mucus film.

3. Mouth terminal. The mouth has well-developed jaws and most fishes have numerous homodont teeth. Paired olfactory sacs may or may not open into the mouth.

4. Bony *operculum*. The operculum protects the fleshy gills that are supported by gill arches.

5. *Swim bladder.* The swim bladder is a hydrostatic organ that may or may not connect to the pharynx. Some fishes have a direct connection of the pharynx to the swim bladder through a *pneumatic duct.* Other fishes lack a connection and have instead a *gas gland* (or *red body*) that extracts gases from the blood to inflate the swim bladder.

6. Two-chambered heart with four pairs of aortic arches. The blood of fishes contains nucleated red blood cells.

7. Sensory portions of the brain are well-developed. The optic lobes are large, there are three pairs of semicircular canals, and 10 pairs of cranial nerves.

8. Separate sexes. Fertilization is usually external. Most bony fishes are oviparous, but some species are ovoviviparous or viviparous.

Class Chondrichthyes

Figure 13.1 Gray smoothhound shark, *Mustelus californicus.*

Figure 13.2 Guitarfish, *Rhina ancylostoma.*

Figure 13.3 Blue spotted stingray, *Taeniura lymma.*

Figure 13.4 Nurse shark, *Ginglymostoma cirratum.*

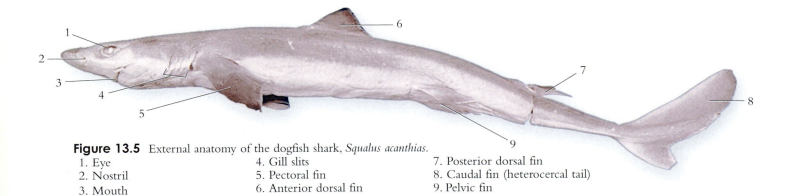

Figure 13.5 External anatomy of the dogfish shark, *Squalus acanthias.*

1. Eye	4. Gill slits	7. Posterior dorsal fin
2. Nostril	5. Pectoral fin	8. Caudal fin (heterocercal tail)
3. Mouth	6. Anterior dorsal fin	9. Pelvic fin

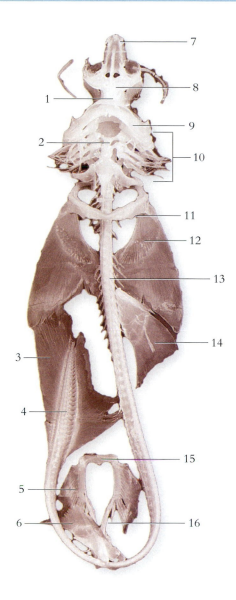

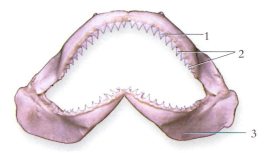

Figure 13.7 Shark jaws.
1. Palatopterygoquadrate cartilage
2. Placoid teeth
3. Meckel's cartilage

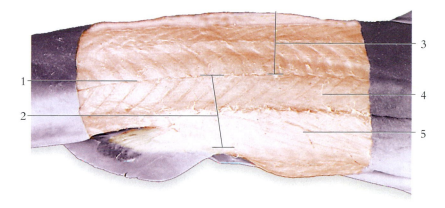

Figure 13.8 Lateral view of the axial musculature of the dogfish shark.
1. Transverse septum
2. Hypaxial myotome portion
3. Epaxial myotome portion
4. Lateral bundle of myotomes
5. Ventral bundle of myotomes

Figure 13.6 Ventral view of the cartilaginous skeleton of a male dogfish shark.
1. Palatopterygoquadrate cartilage (upper jaw)
2. Hypobranchial cartilage
3. Caudal fin
4. Caudal vertebrae
5. Pelvic fin
6. Posterior dorsal fin
7. Rostrum
8. Chondrocranium
9. Meckel's cartilage (lower jaw)
10. Visceral arches
11. Pectoral girdle
12. Pectoral fin
13. Trunk vertebrae
14. Anterior dorsal fin
15. Pelvic girdle
16. Clasper

Figure 13.9 Musculature of the jaw, gills, and pectoral fin of a dogfish shark.
1. 2nd dorsal constrictor
2. Levator of pectoral fin
3. 3rd through 6th ventral constrictors
4. Spiracular muscle
5. Facial nerve (hyomandibular branch)
6. Mandibular adductor
7. 2nd ventral constrictor

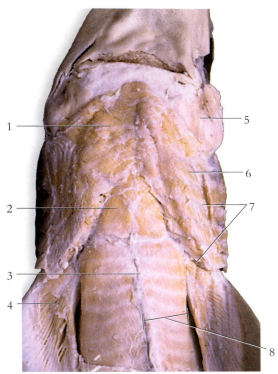

Figure 13.10 Ventral view of the hypobranchial musculature of the dogfish shark.

1. 1st ventral constrictor
2. Common coracoarcual
3. Linea alba
4. Depressor of pectoral fin
5. Mandibular adductor
6. 2nd ventral constrictor
7. 3rd through 6th ventral constrictors
8. Hypaxial muscle

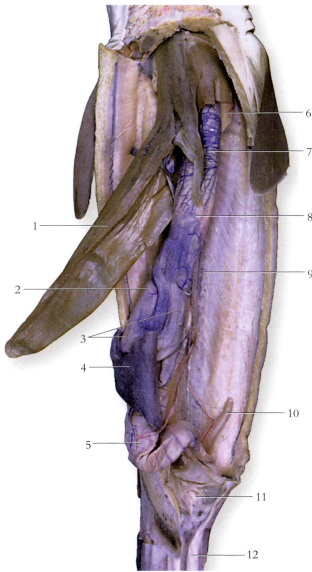

Figure 13.11 Internal anatomy of a male dogfish shark.

1. Right lobe of liver
2. Pyloric sphincter valve
3. Stomach (pyloric region)
4. Spleen
5. Ileum
6. Testis
7. Esophagus
8. Stomach (cardiac region)
9. Kidney
10. Rectal gland
11. Cloaca
12. Clasper

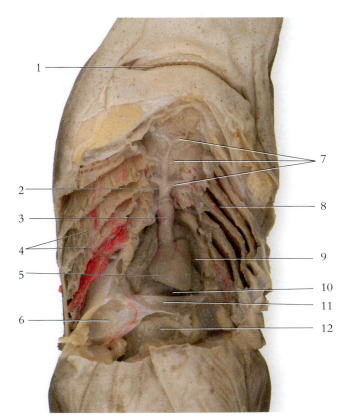

Figure 13.12 Heart, gills, and associated vessels of a dogfish shark.

1. Mouth
2. Ventral aorta
3. Conus ateriosus
4. Gills
5. Ventricle
6. Pectoral girdle (cut)
7. Afferent branchial arteries
8. Gill cleft
9. Pericardial cavity
10. Sinus venosus
11. Transverse septum
12. Liver

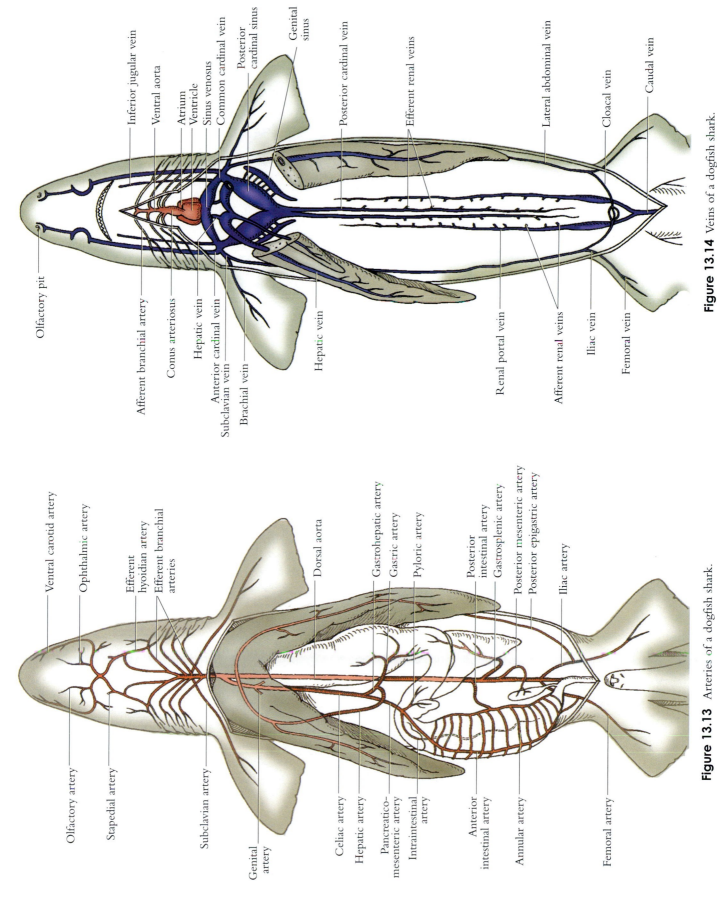

Figure 13.14 Veins of a dogfish shark.

Inferior jugular vein
Ventral aorta
Atrium
Ventricle
Sinus venosus
Common cardinal vein
Posterior cardinal sinus
Genital sinus
Posterior cardinal vein
Efferent renal veins
Lateral abdominal vein
Cloacal vein
Caudal vein

Olfactory pit
Afferent branchial artery
Conus arteriosus
Hepatic vein
Anterior cardinal vein
Subclavian vein
Brachial vein
Hepatic vein
Renal portal vein
Afferent renal veins
Iliac vein
Femoral vein

Figure 13.13 Arteries of a dogfish shark.

Ventral carotid artery
Ophthalmic artery
Efferent hyoidian artery
Efferent branchial arteries
Dorsal aorta
Gastrohepatic artery
Gastric artery
Pyloric artery
Posterior intestinal artery
Gastrosplenic artery
Posterior mesenteric artery
Posterior epigastric artery
Iliac artery

Olfactory artery
Stapedial artery
Subclavian artery
Genital artery
Celiac artery
Hepatic artery
Pancreatico-mesenteric artery
Intraintestinal artery
Anterior intestinal artery
Annular artery
Femoral artery

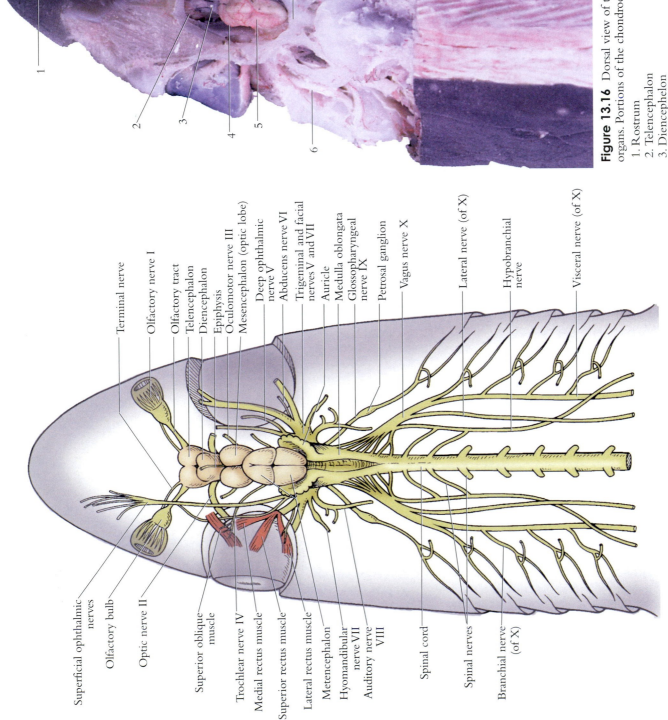

Figure 13.16 Dorsal view of the dogfish brain and sensory organs. Portions of the chondrocranium have been shaved away.

1. Rostrum
2. Telencephalon
3. Diencephalon
4. Mesencephalon
5. Metencephalon
6. Glossopharyngeal nerve
7. Olfactory receptor
8. Eye
9. Chondrocranium
10. Semicircular canal

Superficial ophthalmic nerves
Olfactory bulb
Optic nerve II
Superior oblique muscle
Trochlear nerve IV
Medial rectus muscle
Superior rectus muscle
Lateral rectus muscle
Metencephalon
Hyomandibular nerve VII
Auditory nerve VIII
Spinal cord
Spinal nerves
Branchial nerve (of X)

Terminal nerve
Olfactory nerve I
Olfactory tract
Telencephalon
Diencephalon
Epiphysis
Oculomotor nerve III
Mesencephalon (optic lobe)
Deep ophthalmic nerve V
Abducens nerve VI
Trigeminal and facial nerves V and VII
Auricle
Medulla oblongata
Glossopharyngeal nerve IX
Petrosal ganglion
Vagus nerve X
Lateral nerve (of X)
Hypobranchial nerve
Visceral nerve (of X)

Figure 13.15 Dorsal view of the dogfish brain, cranial nerves, and eye muscles.

Class Osteichthyes

Figure 13.17 Jack mackerel, *Trachurus declivis.*

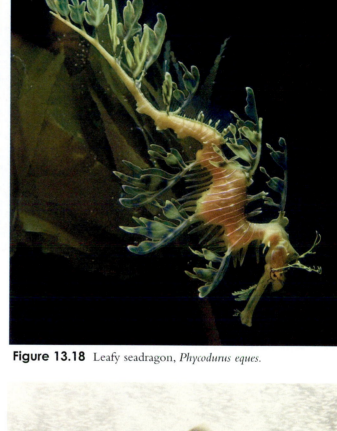

Figure 13.18 Leafy seadragon, *Phycodurus eques.*

Figure 13.19 Blue damsel fish, *Abudefduf cyaneus.*

Figure 13.20 Chum salmon, *Oncorhynchus keta.*

Figure 13.21 Prepared specimen of a sea horse.

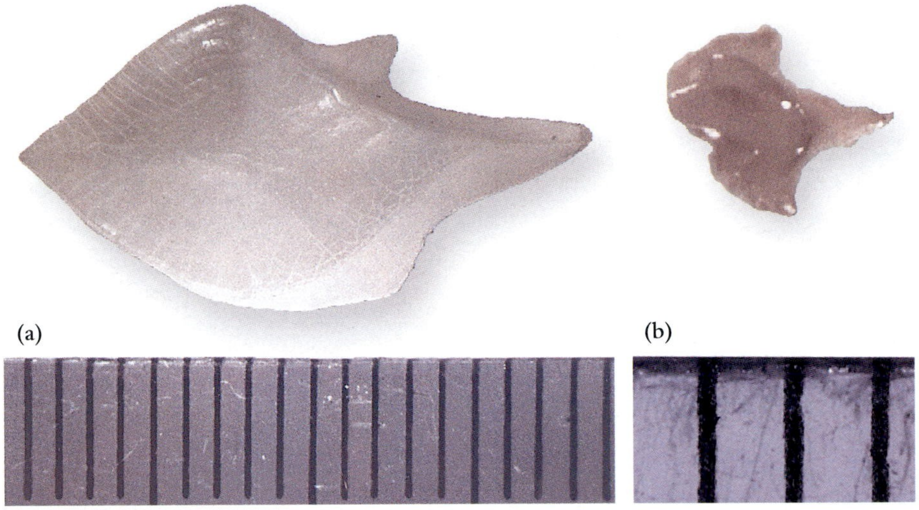

Figure 13.22 Ganoid scales, present in primitive fishes like the gar, are comprised of silvery ganoin on the top surface and bone on the bottom. Two different sizes are shown (a) and (b).

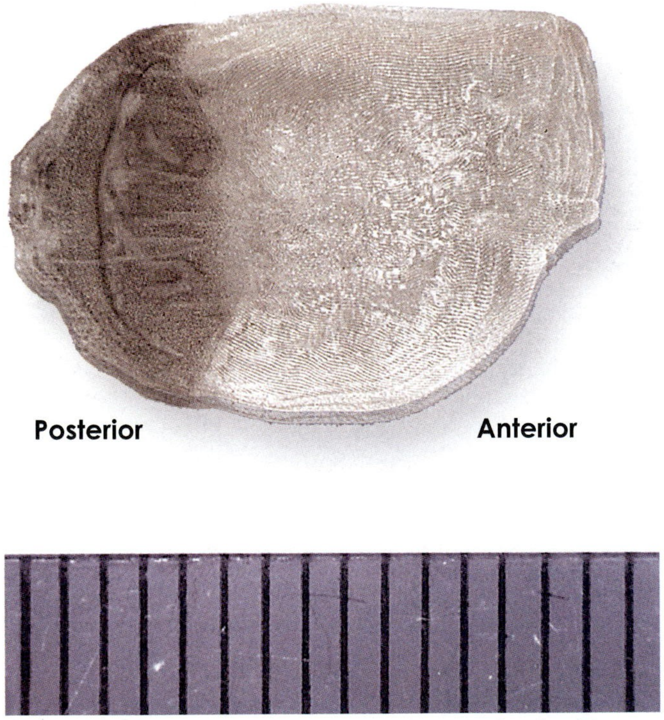

Figure 13.23 Cycloid scales, along with ctenoid scales (Figure 13.24), are found on advanced bony fishes. They are much thinner and more flexible than ganoid scales and overlap each other.

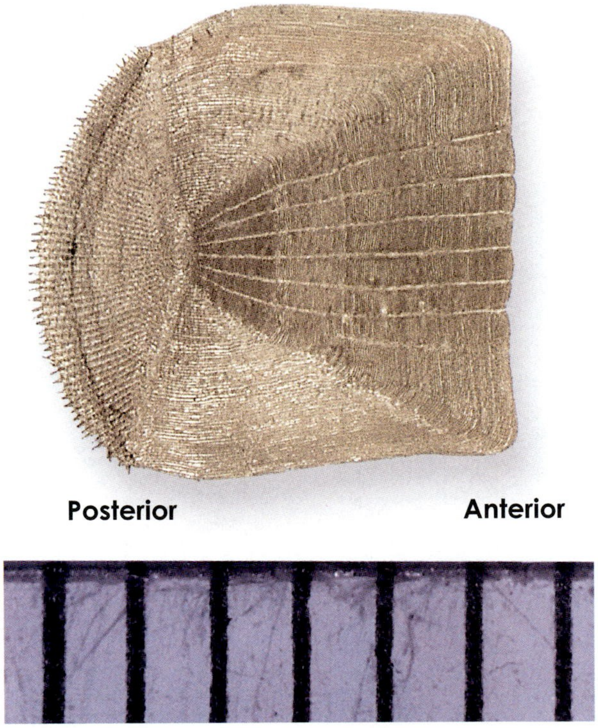

Figure 13.24 Ctenoid scales differ from cycloid scales in that they have comblike ridges on the exposed edge. Which is thought to improve swimming efficiency.

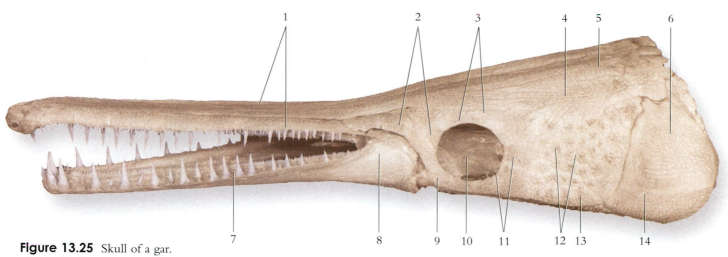

Figure 13.25 Skull of a gar.

1. Maxillaries	4. Pterotic	7. Dentary	10. Orbit	13. Preopercular
2. Preorbitals	5. Parietal	8. Angular	11. Postorbitals	14. Subopercular
3. Supraorbitals	6. Opercular	9. Infraorbital	12. Cheek plates	

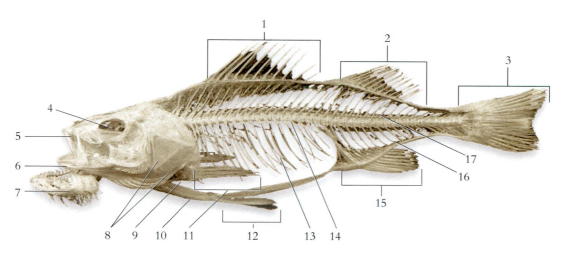

Figure 13.26 Skeleton of a perch.

1. Anterior dorsal fin	6. Dentary	11. Pectoral fin	16. Neural spine
2. Posterior dorsal fin	7. Branchial skeleton	12. Pelvic fin	17. Haemal spine
3. Caudal fin	8. Opercular bones	13. Rib	
4. Orbit	9. Pectoral girdle	14. Vertebral column	
5. Premaxilla	10. Pelvic girdle	15. Anal fin	

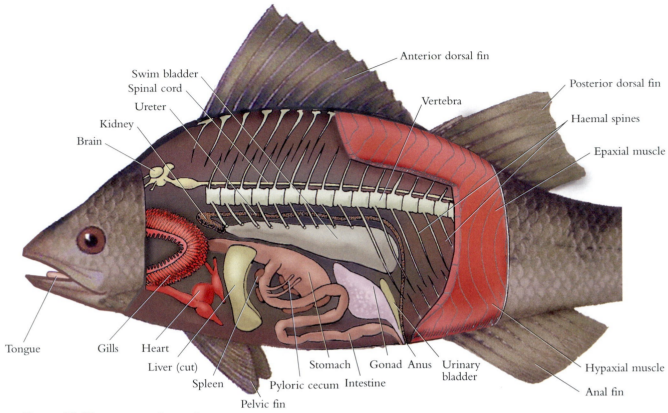

Figure 13.27 Anatomy of a perch.

Figure 13.28 Viscera of a perch.

1. Epaxial muscles
2. Ribs
3. Gill
4. Liver (cut)
5. Heart
6. Pyloric cecum
7. Vertebrae
8. Swim bladder
9. Urinary bladder
10. Gonad
11. Anus
12. Intestine
13. Stomach

There are an estimated 5,540 species of amphibians contained within the class *Amphibia*. Included in this group are caecilians, salamanders, frogs, and toads. Amphibians are vertebrate animals that are transitional in body structure and behavior between aquatic and terrestrial environments. Most larval amphibians are adapted for an aquatic environment. They have gills, elongated tails for swimming, a lateral line system, and a digestive tract adapted for utilization of plant material. Adult amphibians generally occupy a moist terrestrial habitat. Adults breathe by lungs (some have gills), are thin-skinned, and have a specialized three-chambered heart. The amphibian egg is deposited in the water and is fishlike in that it lacks a shell and an amnion.

There are three orders of living amphibians. Order *Gymnophiona* (*Apoda*) includes the estimated 170 species of caecilians. These amphibians lack girdles and appendages, and their tails are short or absent. Most caecilians have *mesodermal* scales. Order *Caudata* includes the estimated 470 species of salamanders. Salamanders have distinct body regions (head, trunk, and tail) and usually two paired appendages. Scales are lacking in salamanders. Order *Anura* (*Salientia*) includes the estimated 4,900 species of frogs and toads. These amphibians have smooth skin (lack scales), fused head and trunk, and two pairs of limbs with the hind limbs adapted for jumping and swimming.

Some of the characteristics of amphibians include:

1. *Variable body forms* — Some amphibians lack limbs, others have limbs adapted for a generalized terrestrial locomotion. Others have highly specialized hind limbs for jumping and swimming. Webbed feet are present in many, with no true claws.

2. *Smooth glandular skin* — The skin glands in many species produce poison. Pigment producing cells, called chromatophores, are common in many amphibians.

3. *Large mouth, small homodont teeth, and many with a protrusile tongue* — All amphibians have paired nostrils opening into the oral cavity.

4. *Respiration through a variety of means* — The moistened skin in many functions as a respiratory surface. Lungs are present in all but a few salamanders. External gills are present in larvae and may persist into adult forms of certain salamanders.

5. *Three-chambered heart* — The amphibian heart has two atria and one ventricle. Blood circulation is through a pulmonary circuitry to the lungs and then back to the heart to be pumped to the body.

6. *Brain is generalized* — Although the sensory portions of the brain may be well developed in some, most amphibians have a simplified brain. Like the fishes, amphibians have ten pairs of cranial nerves.

7. *Paired opisthonephric kidneys and a urinary bladder.*

8. *Separate sexes* — Fertilization is internal by spermatophores in most caecilians and salamanders, and external in most frogs and toads. Most amphibians are oviparous, some are ovoviviparous, and some are viviparous.

Amphibians are extremely important in semiaquatic communities and the general food chain. Most amphibians feed upon insects and they themselves are a source of food for many fishes, birds, and mammals.

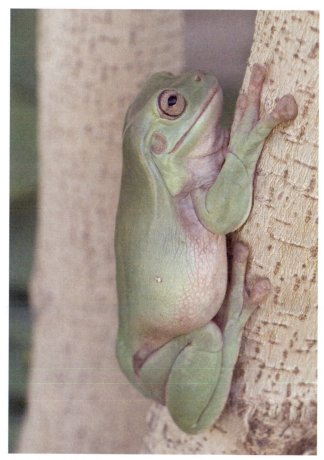

Figure 14.1 White's tree frog, *Litoria caerulea*.

Order Caudata

Figure 14.2 External anatomy of a mud puppy, *Necturus*.

1. Cranium	3. Dentary	5. Forelimb	7. Hind limb
2. Eye	4. External gill	6. Tail	

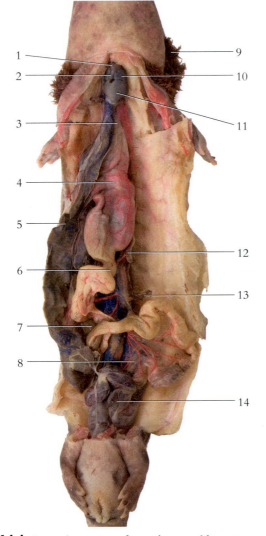

Figure 14.3 Dorsal view of the mud puppy, *Necturus*, skeleton.

1. Frontal bone
2. Squamosal bone
3. Manus (carpal bones, metacarpal bones, phalanges)
4. Ulna
5. Humerus
6. Pes (tarsal bones, metatarsal bones, phalanges)
7. Fibula
8. Pelvic girdle
9. Nasal bone
10. Parietal bone
11. Pectoral girdle
12. Trunk vertebrae
13. Ribs
14. Femur
15. Caudal vertebrae

Figure 14.4 Internal anatomy of a mud puppy, *Necturus*.

1. Bulbus arteriosus
2. Conus arteriosus
3. Esophagus
4. Stomach
5. Liver
6. Duodenum
7. Gallbladder
8. Mesentery
9. External gills
10. Left atrium of heart
11. Ventricle of heart
12. Left lung
13. Pancreas
14. Colon

Order Anura

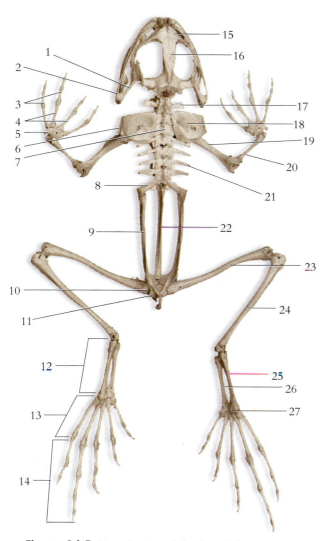

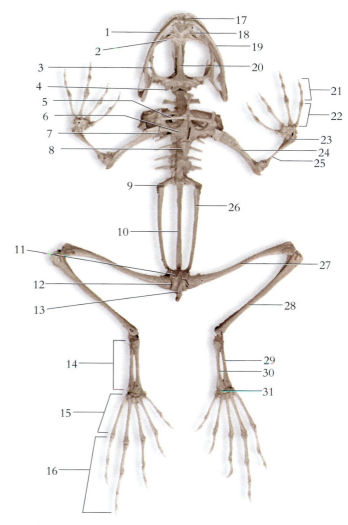

Figure 14.5 Dorsal view of the frog skeleton.

1. Squamosal bone
2. Quadratojugal bone
3. Phalanges of digits
4. Metacarpal bones
5. Carpal bones
6. Scapula
7. Vertebra
8. Transverse process of sacral (ninth) vertebra
9. Ilium
10. Acetabulum
11. Ischium
12. Tarsal bones
13. Metatarsal bones
14. Phalanges of digits
15. Nasal bone
16. Frontoparietal bone
17. Transverse process
18. Suprascapula
19. Humerus
20. Radioulna
21. Transverse process
22. Urostyle
23. Femur
24. Tibiofibula
25. Fibulare (calcaneum)
26. Tibiale (astragalus)
27. Distal tarsal bones

Figure 14.6 Ventral view of the frog skeleton.

1. Maxilla
2. Palatine
3. Pterygoid bone
4. Exoccipital bone
5. Clavicle
6. Coracoid
7. Glenoid fossa
8. Sternum
9. Transverse process of sacral (ninth) vertebra
10. Urostyle
11. Pubis
12. Acetabulum
13. Ischium
14. Tarsal bones
15. Metatarsal bones
16. Phalanges of digits
17. Premaxilla
18. Vomer
19. Dentary
20. Parasphenoid bone
21. Phalanges of digits
22. Metacarpal bones
23. Carpal bones
24. Humerus
25. Radioulna
26. Ilium
27. Femur
28. Tibiofibula
29. Fibulare (calcaneum)
30. Tibiale (astragalus)
31. Distal tarsal bones

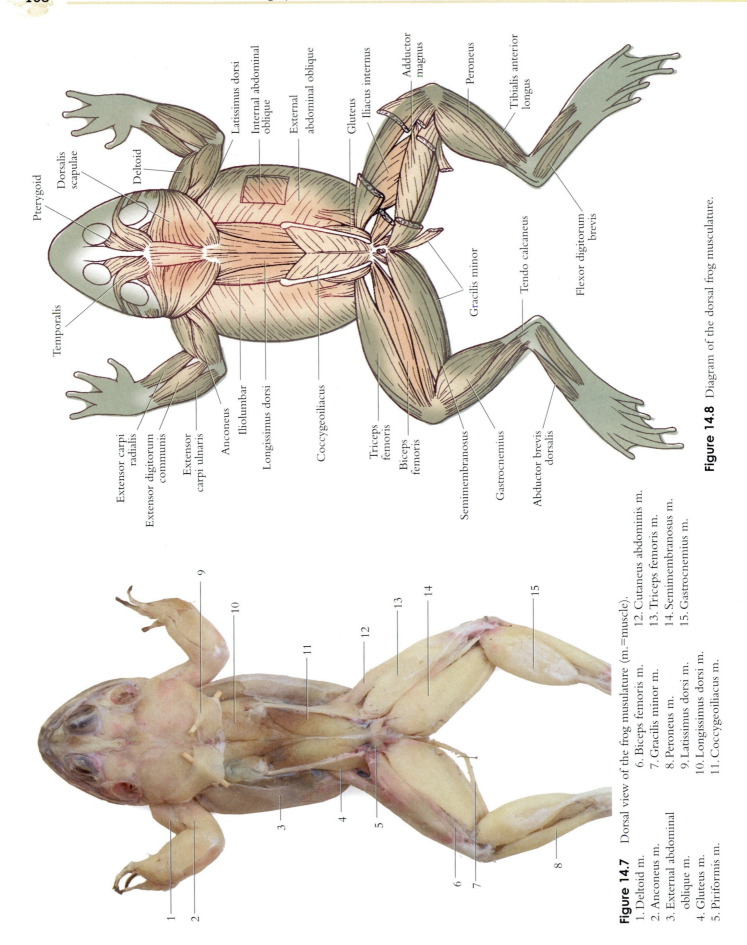

Figure 14.8 Diagram of the dorsal frog musculature.

Figure 14.7 Dorsal view of the frog musulature (m.=muscle).

1. Deltoid m.	6. Biceps femoris m.	12. Cutaneus abdominis m.
2. Anconeus m.	7. Gracilis minor m.	13. Triceps femoris m.
3. External abdominal	8. Peroneus m.	14. Semimembranosus m.
oblique m.	9. Latissimus dorsi m.	15. Gastrocnemius m.
4. Gluteus m.	10. Longissimus dorsi m.	
5. Piriformis m.	11. Coccygeoiliacus m.	

Labels on Figure 14.8 (diagram):

Pterygoid, Dorsalis scapulae, Deltoid, Latissimus dorsi, Internal abdominal oblique, External abdominal oblique, Gluteus, Iliacus internus, Adductor magnus, Peroneus, Tibialis anterior longus, Temporalis, Extensor carpi radialis, Extensor digitorum communis, Extensor carpi ulnaris, Anconeus, Iliolumbar, Longissimus dorsi, Coccygeoiliacus, Triceps femoris, Biceps femoris, Semimembranosus, Gastrocnemius, Abductor brevis dorsalis, Gracilis minor, Tendo calcaneus, Flexor digitorum brevis

Figure 14.9 Dorsal view of the leg muscles of a frog (m.=muscle).

1. Gluteus m.
2. Cutaneus abdominis m.
3. Piriformis m.
4. Semimembranosus m.
5. Gracilis minor m.
6. Peroneus m.
7. Coccygeoiliacus m.
8. Triceps femoris m. (cut)
9. Iliacus internus m.
10. Biceps femoris m.
11. Adductor magnus m.
12. Semitendinosus m .
13. Gastrocnemius m.

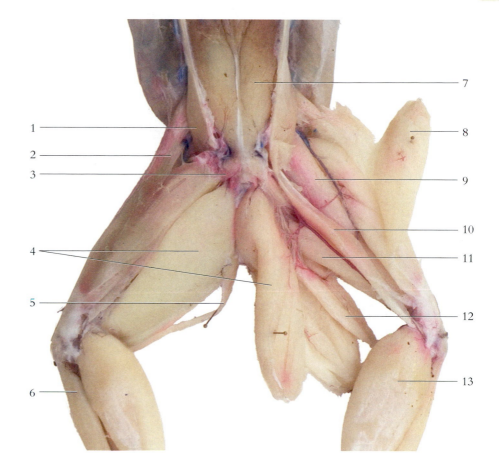

Figure 14.10 A ventral view of the leg muscles of a frog (m.=muscle).

1. External abdominal oblique m.
2. Triceps femoris m.
3. Adductor longus m.
4. Adductor magnus m.
5. Semitendinosus m.
6. Semimembranosus m.
7. Gastrocnemius m.
8. Tibialis posterior m.
9. Rectus abdominis m.
10. Sartorius m.
11. Gracilis major m.
12. Gracilis minor m.
13. Extensor cruris m.
14. Tibialis anterior longus m.
15. Tibialis anterior brevis m.

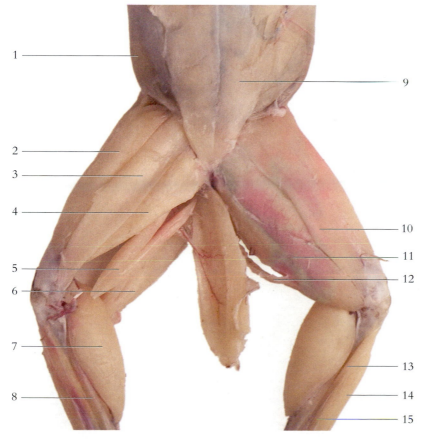

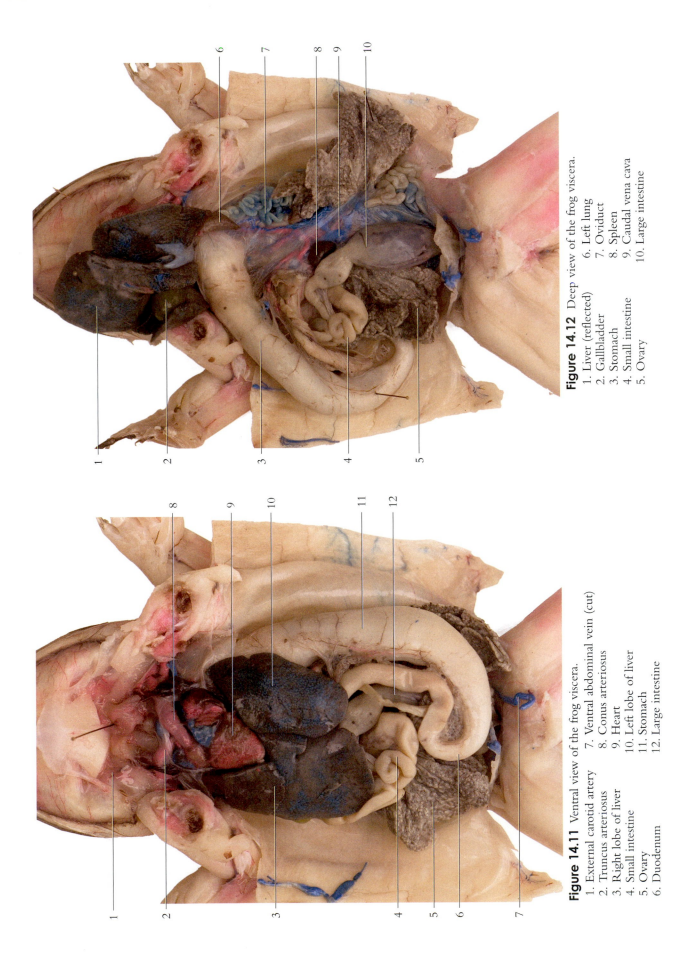

Figure 14.12 Deep view of the frog viscera.

1. Liver (reflected)
2. Gallbladder
3. Stomach
4. Small intestine
5. Ovary
6. Left lung
7. Oviduct
8. Spleen
9. Caudal vena cava
10. Large intestine

Figure 14.11 Ventral view of the frog viscera.

1. External carotid artery
2. Truncus arteriosus
3. Right lobe of liver
4. Small intestine
5. Ovary
6. Duodenum
7. Ventral abdominal vein (cut)
8. Conus arteriosus
9. Heart
10. Left lobe of liver
11. Stomach
12. Large intestine

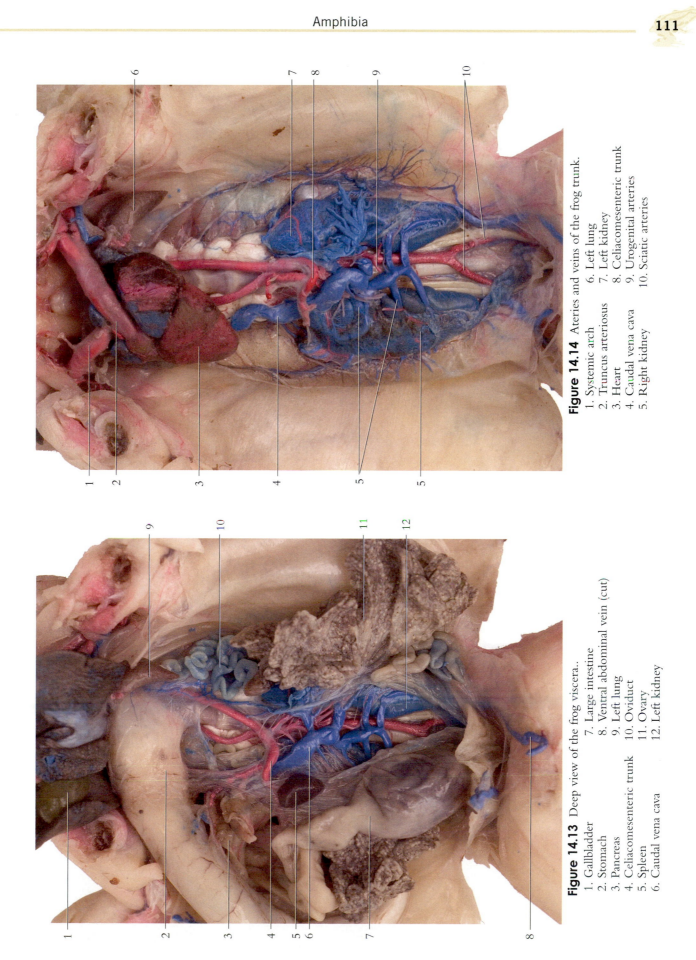

Figure 14.14 Ateries and veins of the frog trunk.

1. Systemic arch
2. Truncus arteriosus
3. Heart
4. Caudal vena cava
5. Right kidney
6. Left lung
7. Left kidney
8. Celiacomesenteric trunk
9. Urogenital arteries
10. Sciatic arteries

Figure 14.13 Deep view of the frog viscera..

1. Gallbladder
2. Stomach
3. Pancreas
4. Celiacomesenteric trunk
5. Spleen
6. Caudal vena cava
7. Large intestine
8. Ventral abdominal vein (cut)
9. Left lung
10. Oviduct
11. Ovary
12. Left kidney

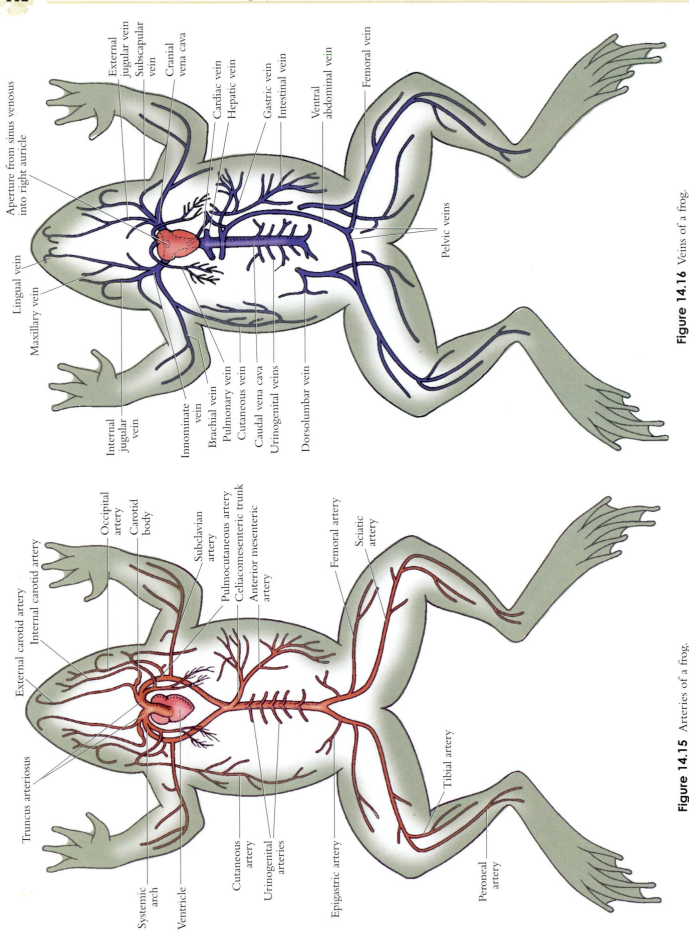

Figure 14.16 Veins of a frog.

Figure 14.15 Arteries of a frog.

Reptiles are chordate animals within the class *Reptilia*. There are about 7,500 species of reptiles represented in the common forms of animals known as turtles, tuataras, lizards snakes, crocodiles, and alligators. Generally, reptiles are considered ectothermous, or poikilothermous (cold-blooded), although some maintain a fairly consistent endothermous (warm-blooded) condition through behavioral adaptations.

Although most of the species of reptiles that once lived are now extinct, they were highly successful in terms of diversity and duration of existence. Reptiles dominated the earth for millions of years during the Mesozoic era. Birds and mammals are descendants of stem reptiles. Living reptiles are classified into three subclasses and three or four (depending upon classification scheme) orders. Subclass *Anapsida* includes the reptiles that lack a temporal opening in the skull. Order *Testudinata* within the subclass Anapsida includes the turtles (aquatic forms) and tortoises (terrestrial forms). Subclass *Lepidosauria* includes the reptiles that have two temporal openings in the skull. Order *Squamata* within the subclass Lepidosauria includes the lizards, snakes, and tuataras. Some classifications assign the tuatara (*Sphenodon*) to its own order, *Rynchocephalia*. Subclass *Archosauria* includes the ruling Mesozoic reptiles (dinosaurs and pterosaurs). Order *Crocodilia* within the subclass Archosauria includes the crocodiles, alligators, caimans, and gavials.

Except for a few species of sea snakes and sea turtles, reptiles complete their entire life cycle on land and are considered terrestrial vertebrates. Adaptations that permit utilization of non-aqueous habitats include *keratinized scales* that protect the animal from desiccation, *internal fertilization*, and *amniotic eggs* protected by leathery shells. *Keratin* is a protein within scales that contributes to their strength and waterproofing capabilities. Internal fertilization prevents desiccation of the sperm and requires a male intromittent organ. Male turtles and tortoises, for example, have a *penis* and male snakes and lizards have a *hemipene* to transport ejaculated sperm to the female cloaca. The *amnion* is a thin embryonic membrane which insures that development is within the homogeneous and protective environment of *amniotic fluid*. Because the embryonic development of reptiles, birds, and mammals is within amniotic fluid, these groups of vertebrate animals are sometimes referred to as *amniotes*. The *anamniotes* are the cyclostomes, fishes, and amphibians, or the vertebrates that lack an amnion and amniotic fluid.

Some of the characteristics of reptiles include:

1. A dry scaly skin that is keratinized to protect against desiccation.

2. Internal fertilization usually involving an intromittent, or copulatory, organ to maximize the survival of the gametes.

3. Development of the shelled amniotic egg, in which the embryo forms in a protective fluid environment.

4. The lack of a larval stage of development.

5. Except for snakes and some lizards, the body is supported by girdles and appendages that permit effective terrestrial locomotion. Snakes have secondarily lost their limbs through a selective evolutionary process.

6. *Single occipital condyle* — The articulation between the skull and the atlas vertebra of the vertebral column is through a single bony protuberance. Reptiles and birds have this arrangement, whereas amphibians and mammals have double occipital condyles.

7. *Poikilothermous* — Of the vertebrates, only the birds and mammals are homeothermous (metabolically generate a consistent internal body temperature), or endothermous. All other vertebrates are poikilothermous (or ectothermous), or are dependent upon environmental factors to insure a functional body temperature. Some reptiles, however, are able to maintain a relatively consistent internal body temperature through behavioral adaptations.

8. Enlarged sensory perception portions of the brain and twelve pairs of cranial nerves.

Figure 15.1 African gaboon viper, *Bitis gabonica rhinoceros*, has cryptic coloration to blend with its background.

Subclass Anapsida   Order Chelonia

Figure 15.2 Hawaiian green turtle, *Chelonia mydas,* (a) and the desert tortoise, *Gopherus agassizii,* (b) are just two of the many members of order chelonia that are threatened or endangered.

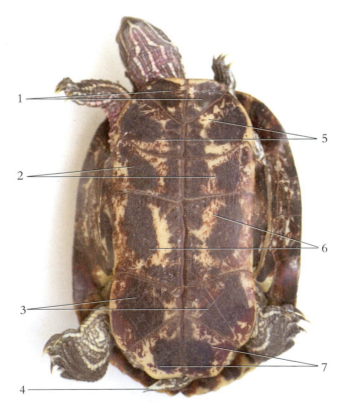

Figure 15.3 Dorsal view of a turtle.

1. Eye	5. Nostril
2. Pentadactyl foot	6. Head
3. Vertebral scales	7. Nuchal scale
4. Marginal scales	8. Costal scales
(encircle the carapace)	

Figure 15.4 Ventral view of a turtle.

1. Gular scales	5. Humeral scales
2. Pectoral scales	6. Abdominal scales
3. Femoral scales	7. Anal scales
4. Tail	

Figure 15.5 Skull of a turtle.

1. Parietal bone
2. Supraoccipital bone
3. Postorbital bone
4. Jugal bone
5. Quadratojugal bone
6. Exoccipital bone
7. Quadrate bone
8. Supraangular bone
9. Articular bone
10. Angular bone
11. Frontal bone
12. Prefrontal bone
13. Palatine bone
14. Premaxilla
15. Maxilla
16. Beak
17. Dentary

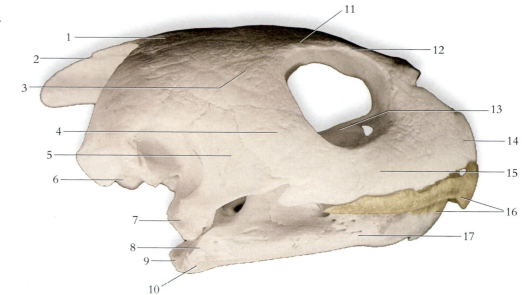

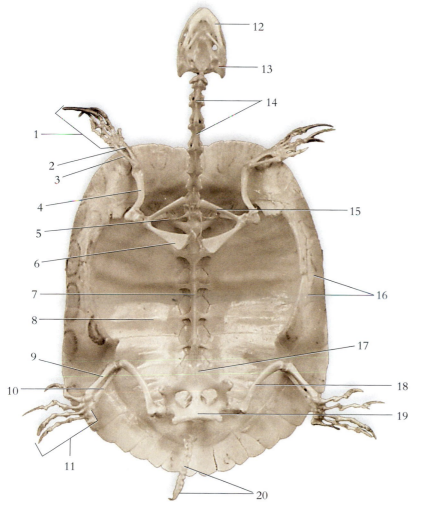

Figure 15.6 Skeleton of a turtle.
(The plastron is removed.)

1. Manus (carpal bones, metacarpal bones, phalanges)
2. Radius
3. Ulna
4. Humerus
5. Procoracoid
6. Scapula
7. Vertebra
8. Rib
9. Tibia
10. Fibula
11. Pes (tarsal bones, metatarsal bones, phalanges)
12. Dentary
13. Articular
14. Cervical vertebrae
15. Acromion process
16. Dermal plate of carapace
17. Pubis
18. Femur
19. Ischium
20. Caudal vertebrae

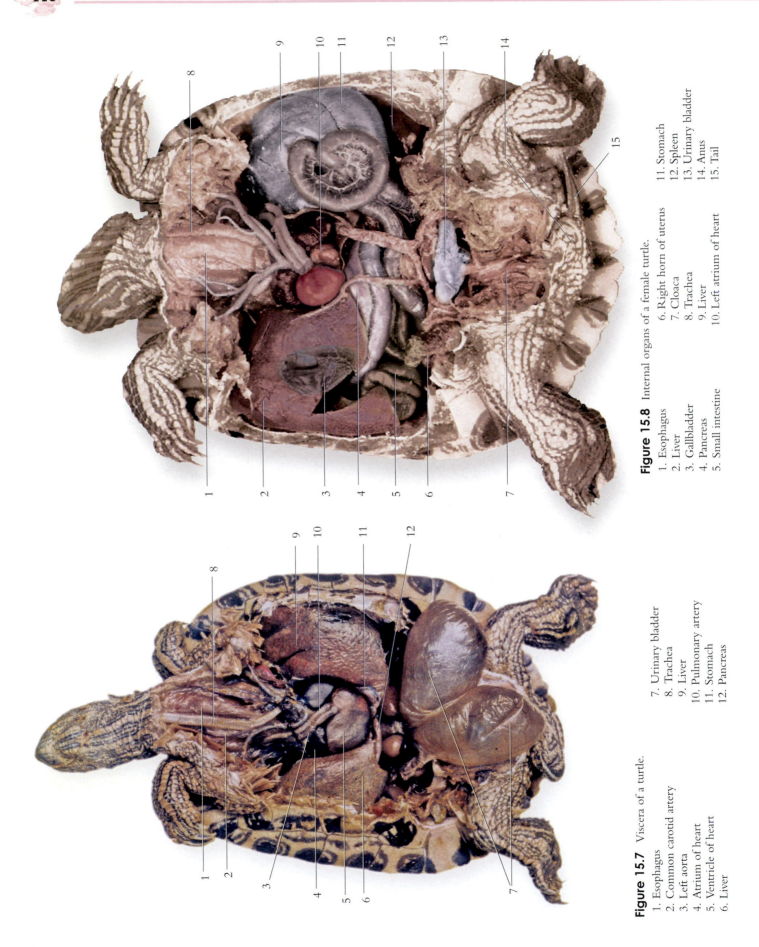

Figure 15.8 Internal organs of a female turtle.

1. Esophagus
2. Liver
3. Gallbladder
4. Pancreas
5. Small intestine
6. Right horn of uterus
7. Cloaca
8. Trachea
9. Liver
10. Left atrium of heart
11. Stomach
12. Spleen
13. Urinary bladder
14. Anus
15. Tail

Figure 15.7 Viscera of a turtle.

1. Esophagus
2. Common carotid artery
3. Left aorta
4. Atrium of heart
5. Ventricle of heart
6. Liver
7. Urinary bladder
8. Trachea
9. Liver
10. Pulmonary artery
11. Stomach
12. Pancreas

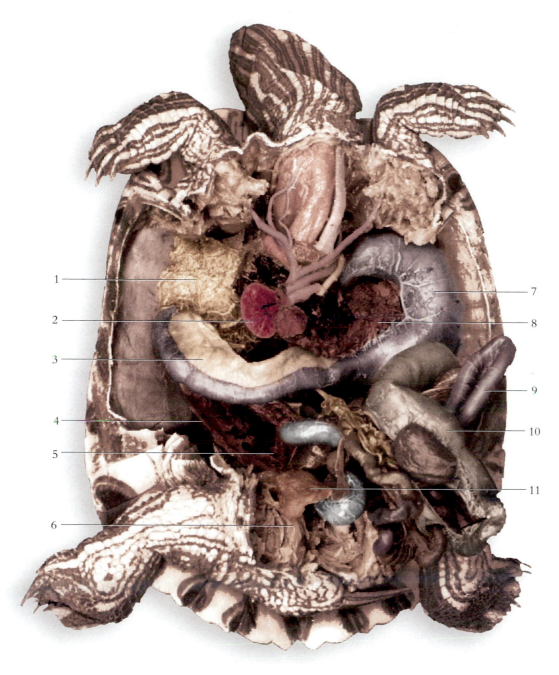

Figure 15.9 Internal organs of a female turtle with the liver removed to show the deeper organs.

1. Right lung
2. Ventricle of heart
3. Pancreas
4. Ureter
5. Ovary
6. Cloaca
7. Stomach
8. Liver (cut)
9. Small intestine
10. Large intestine
11. Urinary bladder

Subclass Lepidosauria   Order Squamata

Figure 15.10 Side-blotched lizard, *Uta stansburiana*, in breeding color.

Figure 15.11 Sidewinder rattlesnake, *Crotalus cerastes*.

Figure 15.12 California king snake, *Lampropeltis gentulus*.

Figure 15.13 European glass lizard, *Ophisaurus apodus*, is a legless lizard.

(a)

(b)

Figure 15.14 Comparison of the external anatomy of the head of (a) the savannah monitor lizard (*Varanus exanthematicus*) and (b) the green tree python (*Morelia viridis*). Lizards have eyelids and external ears, whereas, these structures are lacking in snakes. (Notice the heat pit receptors, which are characteristic of pythons and pit vipers and are adaptive for predation on warm-blooded vertebrates.)

1. Eyelids   2. External ear   3. Heat pits

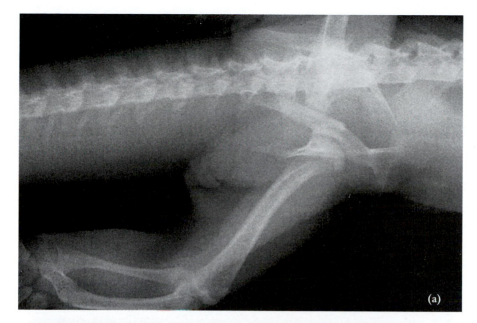

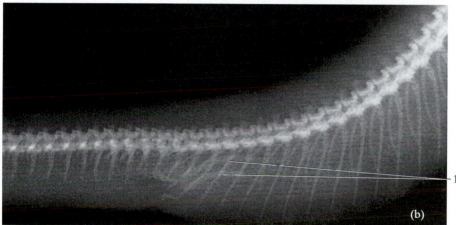

Figure 15.15 Radiograph of the pelvic region of a savannah monitor (a) showing a highly developed limb. Compare this to the radiograph of the pelvic region of a boa (b) showing the vestigeal pelvic girdle.
  1. Vestigeal pelvic girdle

Figure 15.16 Radiograph of the skeleton of the Argus monitor, *Varanus panoptes*.

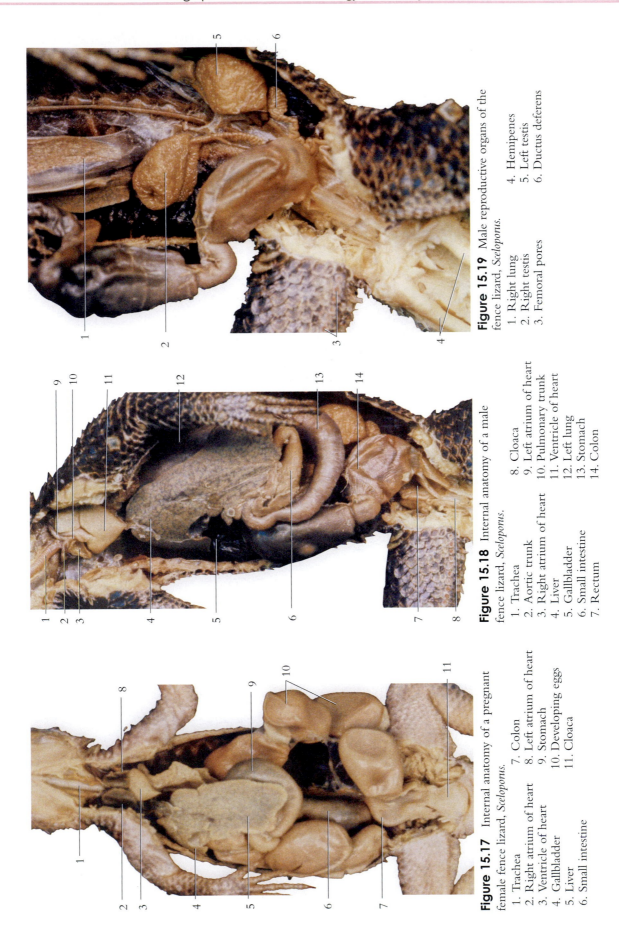

Figure 15.19 Male reproductive organs of the fence lizard, *Sceloporus.*

1. Right lung
2. Right testis
3. Femoral pores
4. Hemipenes
5. Left testis
6. Ductus deferens

Figure 15.18 Internal anatomy of a male fence lizard, *Sceloporus.*

1. Trachea
2. Aortic trunk
3. Right atrium of heart
4. Liver
5. Gallbladder
6. Small intestine
7. Rectum
8. Cloaca
9. Left atrium of heart
10. Pulmonary trunk
11. Ventricle of heart
12. Left lung
13. Stomach
14. Colon

Figure 15.17 Internal anatomy of a pregnant female fence lizard, *Sceloporus.*

1. Trachea
2. Right atrium of heart
3. Ventricle of heart
4. Gallbladder
5. Liver
6. Small intestine
7. Colon
8. Left atrium of heart
9. Stomach
10. Developing eggs
11. Cloaca

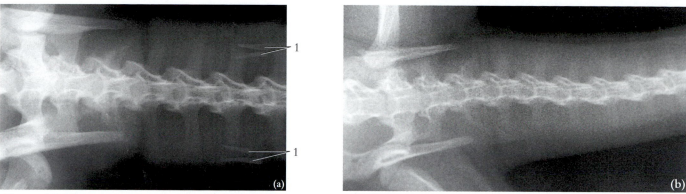

Figure 15.20 Male lizards and snakes have hemipenes as copulatory organs. The hemipene seen in a radiograph of a male (a) crocodile monitor, *Varanus salvadorii*. As seen in a radiograph, a female (b) lacks a hemipene. The female cloaca is the receptacle of the everted male hemipene during copulation.
  1. Sheaths of hemipenes

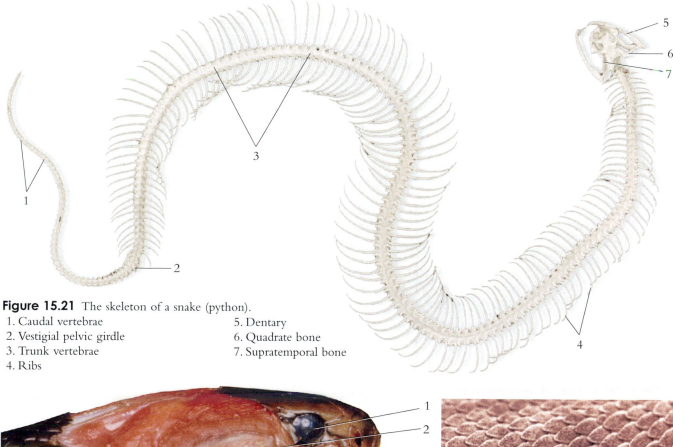

Figure 15.21 The skeleton of a snake (python).
  1. Caudal vertebrae
  2. Vestigial pelvic girdle
  3. Trunk vertebrae
  4. Ribs
  5. Dentary
  6. Quadrate bone
  7. Supratemporal bone

Figure 15.22 Oral region of a cobra, *Naja*, showing its poison gland, which is a modified salivary gland; and its fang, which is a modified maxillary tooth.
  1. Eye
  2. Poison gland
  3. Fang sheath
  4. Fang

Figure 15.23 Overlapping arrangement of snake scales.

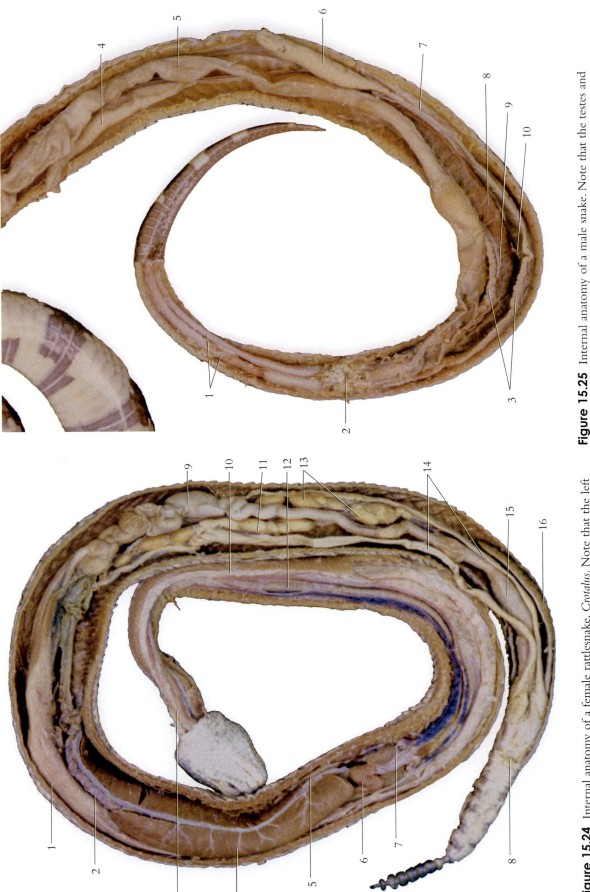

Figure 15.24 Internal anatomy of a female rattlesnake, *Crotalus*. Note that the left lung of snakes is either rudimentary or absent.

1. Stomach
2. Pancreas
3. Trachea
4. Liver
5. Right lung
6. Heart
7. Thyroid gland
8. Anus
9. Intestine
10. Esophagus
11. Right ovary
12. Trachea
13. Left ovary
14. Oviducts
15. Rectum
16. Left kidney

Figure 15.25 Internal anatomy of a male snake. Note that the testes and kidneys are staggered.

1. Hemipenes
2. Anus
3. Ductus deferens
4. Right testis
5. Intestine
6. Left testis
7. Ductus deferens
8. Right kidney
9. Ureter
10. Left kidney

Birds are homeothermous (warm-blooded) chordate animals within the class Aves. There are about 8,600 species of birds. Class Aves is divided into two subclasses: *Archaeornithes* and *Neornithes*. Subclass *Archaeornithes* includes the fossil, *Archaeopteryx*, and subclass *Neornithes* includes all other birds, fossil and modern.

Birds are closely related to reptiles and seem to have evolved from a basic archosaurian stock of quadrupedal animals in the Jurassic period some 140 million years ago. Crocodiles and alligators are also archosaurs and are the closest living relatives of birds. Crocodiles and alligators, however, have specializations for an amphibious life whereas birds have specializations for flight and bipedalism (support on two appendages).

Some of the characteristics of birds include:

1. *Feathers* — Derived from the epidermis of the skin, feathers provide protection, insulation necessary for homeothermia, and lightweight body surface for flight.

2. *Beaks* — Derived from the epidermis of the skin, beaks are highly adapted for a variety of functions, including obtaining food, building nests, preening, defense, and mating. Birds lack teeth and the beaks of many birds are analogous structures.

3. *Thin skin* — The skin of birds is thin, vascular, and has few glands. Oil-secreting glands line the outer ear canal and encircle the vent of the cloaca in some birds. The uropygial gland at the base of the tail is also an oil-secreting gland that is used in preening and is especially well developed in aquatic birds.

4. *Hollow bones* — The sternum, vertebrae, and several of the long bones of birds are pneumatic, meaning that they have a hollow, air-filled core. Pneumatic bones contribute to the skeletal system of a bird being lightweight and yet strong and durable.

5. *Fusion of bones* — A reduction in the number of bones also contributes to the light weight body of a bird, while still providing structural support. Other bones have specializations to accommodate flight. The bones of the forelimbs, for example, are adapted for flight, and the ribs have uncinate processes which strengthen the ribcage and provide an increased surface area for muscle attachment.

6. *Sclerotic bones* — A ring of bones within the anterior portion of the eyeball of birds and certain reptiles provides structural support.

7. *Single occipital condyle* — The articulation between the skull and the atlas vertebra of the vertebral column is through a single occipital condyle, rather than a double occipital condyle as is the arrangement in mammals.

8. *Skeletal muscles positioned primarily upon the torso* — The large and powerful flight muscles are positioned upon the sternum and rib cage along the ventral side of the thorax. In this position, these heavy muscles provide ballast to the bird during flight. The muscles that move the distal portions of the appendages are positioned proximally and act upon their insertion attachment points by long, lightweight tendons.

9. *Four-chambered heart with a right aortic arch* — Although mammals also have a four-chambered heart, their aortic arch extends to the left rather than to the right. A four-chambered heart and separate pulmonary (to the lungs) and systemic (to the body) blood flow are important adaptations for homeothermia.

10. *Single functional ovary* — Although present during embryonic development, in most birds the right ovary and oviduct become vestigial so that only the left genital structures are functional.

11. *Oviparous* — Birds have internal fertilization and are egg-laying, or oviparous.

12. *Air sacs* — A series of air sacs (generally nine) is situated in various parts of the body of a bird and functions principally as air reservoirs. The air sacs are connected to the bronchi of the lungs.

13. *Well-developed brain* — Although non-convoluted, the large cerebrum of the avian brain enables an array of intricate instinctive behaviors, including courtship, nest building, and migration.

Figure 16.1 Hooded merganser, *Lophodytes cucullatus*, is the smallest of the North American mergansers.

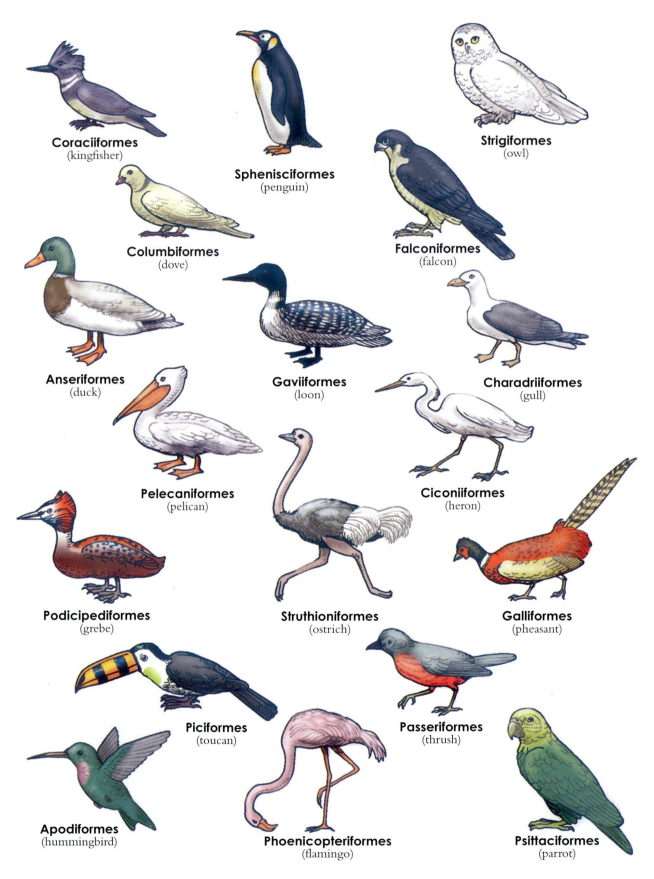

Coraciiformes
(kingfisher)

Sphenisciformes
(penguin)

Strigiformes
(owl)

Columbiformes
(dove)

Falconiformes
(falcon)

Anseriformes
(duck)

Gaviiformes
(loon)

Charadriiformes
(gull)

Pelecaniformes
(pelican)

Ciconiiformes
(heron)

Podicipediformes
(grebe)

Struthioniformes
(ostrich)

Galliformes
(pheasant)

Piciformes
(toucan)

Passeriformes
(thrush)

Apodiformes
(hummingbird)

Phoenicopteriformes
(flamingo)

Psittaciformes
(parrot)

Figure 16.2 Representatives from some of the orders of birds.

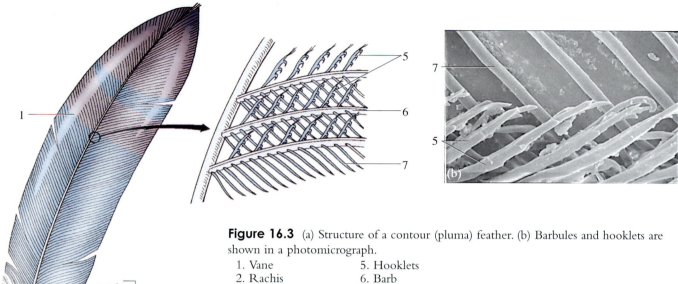

Figure 16.3 (a) Structure of a contour (pluma) feather. (b) Barbules and hooklets are shown in a photomicrograph.

1. Vane
2. Rachis
3. Calamus
4. Shaft
5. Hooklets
6. Barb
7. Barbule

Figure 16.4 Pluma (*pl. plumae*) are contour feathers that include the quill feathers and flight feathers. Quill feathers cover most of the body. Flight feathers are confined to the wings as remiges and to the tail as retrices. A portion of a remige is shown in this micrograph.

Figure 16.5 Plumula (*pl. plumulae*) is a soft insulating feather that lacks a rachis and hooklets. Also called down feathers, plumulae are abundant on the breast and abdomen of a bird. Plumulae are shown in this micrograph.

Figure 16.6 Filoplume (*pl. filoplumes*) is a hairlike feather consisting of a single rachis and either no barbs or just a few at its distal tip. Filoplumes are scattered throughout the plumae. A filoplume is shown in this micrograph.

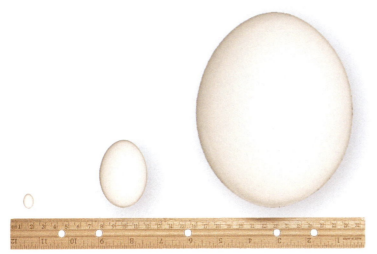

Figure 16.7 Size comparison of bird eggs; (a) a hummingbird, (b) a chicken, and (c) an ostrich.

Figure 16.8 Great blue heron, *Ardea herodias.*

Figure 16.9 Rock dove, *Columba.*

Figure 16.10 California quail, *Callipepla californica.*

Figure 16.11 Mallard duck, *Anas platyrhnchos.*

Figure 16.12 Nighthawk, *Chordeiles minor.*

Figure 16.15 Flocking to a roosting site is an important social behavior in many species of birds.

Figure 16.13 American kestrel, *Falco sparverius.*

Figure 16.14 Broad-tailed humming-bird (female), *Selasphorus platycercus.*

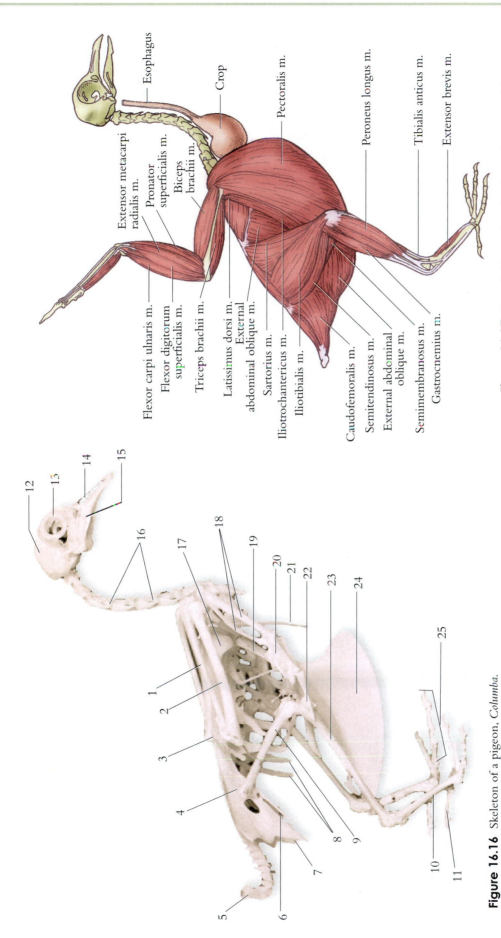

Figure 16.17 Superficial muscles of a pigeon. (m. = muscle)

Labels for Figure 16.17:
Esophagus
Crop
Pectoralis m.
Peroneus longus m.
Tibialis anticus m.
Extensor brevis m.
Extensor metacarpi radialis m.
Pronator superficialis m.
Biceps brachii m.
Flexor carpi ulnaris m.
Flexor digitorum superficialis m.
Triceps brachii m.
Latissimus dorsi m.
External abdominal oblique m.
Sartorius m.
Iliotrochantericus m.
Iliotibialis m.
Caudofemoralis m.
Semitendinosus m.
External abdominal oblique m.
Semimembranosus m.
Gastrocnemius m.

Figure 16.16 Skeleton of a pigeon, *Columba.*

1. Radius
2. Ulna
3. Scapula
4. Ilium
5. Pygostyle
6. Pubis
7. Ischium
8. Ribs
9. Femur
10. Tarsometatarsal bone
11. Digit 1
12. Cranium
13. Sclerotic bone
14. Premaxilla
15. Dentary
16. Cervical vertebrae
17. Humerus
18. Carpometacarpal bones
19. Coracoid bone
20. Phalanges
21. Furcula
22. Phalanx of third digit
23. Tibiotarsal bone
24. Keel of sternum
25. Phalanges

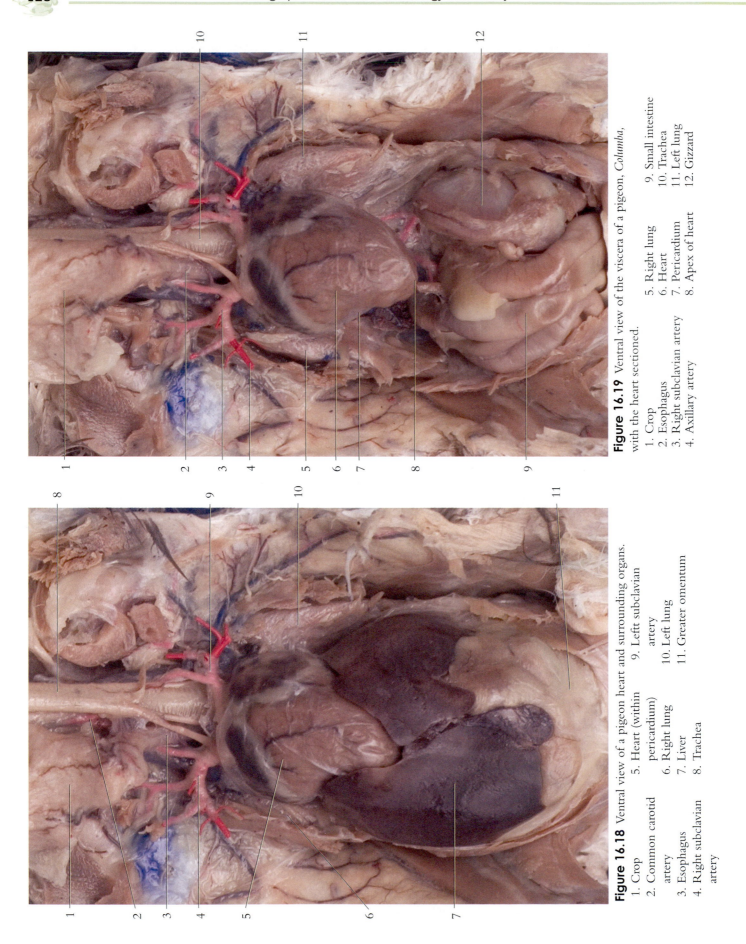

Figure 16.18 Ventral view of a pigeon heart and surrounding organs.

1. Crop
2. Common carotid artery
3. Esophagus
4. Right subclavian artery
5. Heart (within pericardium)
6. Right lung
7. Liver
8. Trachea
9. Left subclavian artery
10. Left lung
11. Greater omentum

Figure 16.19 Ventral view of the viscera of a pigeon, *Columba*, with the heart sectioned.

1. Crop
2. Esophagus
3. Right subclavian artery
4. Axillary artery
5. Right lung
6. Heart
7. Pericardium
8. Apex of heart
9. Small intestine
10. Trachea
11. Left lung
12. Gizzard

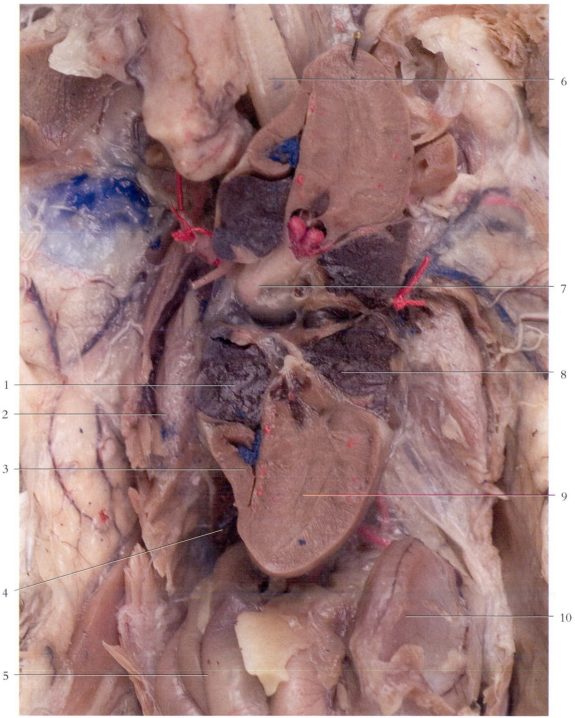

Figure 16.20 Viscera of a pigeon with the heart sectioned.

1. Right atrium
2. Right lung
3. Right ventricle
4. Liver (cut)

5. Small intestine
6. Trachea
7. Aortic arch
8. Left atrium

9. Left ventricle
10. Gizzard

Chapter 17
Mammalia

Mammals are homeothermous (warm-blooded) chordate animals within the class *Mammalia*. There are about 4,100 species of mammals. Class *Mammalia* is divided into two subclasses: *Prototheria* and *Theria*. Subclass *Prototheria* includes the *monotremes*, or egg-laying mammals. Monotremes are the only mammals that have a cloaca. Subclass *Theria* includes infraclass *Metatheria* (pouched-mammals) with one order, *Marsupalia*, and infraclass *Eutheria* (placental-mammals) with 19 orders. Monotremes and marsupials are primitive mammals that have certain characteristics (teeth, skeletal) that link them to the fossil therapsid reptiles of the early Mesozoic era.

The advanced placental mammals give birth to relatively well-developed young. Some of the characteristics of mammals include:

1. *Hair* — Derived from the epidermis of the skin, hair provides protection and insulation necessary for homeothermia. Some mammals are sparsely haired, including marine mammals and humans.

2. *Mammary glands* — Modified sweat glands, the mammary glands produce milk upon parturition (birth) for suckling the young.

3. *Four-chambered heart with a left aortic arch* — Although birds also have a four-chambered heart, their aortic arch extends to the right rather than to the left. A four-chambered heart and separate pulmonary (to the lungs) and systemic (to the body) blood flow are important adaptations for homeothermia.

4. *Well-developed brain with a convoluted cerebrum* — Not only is the brain of mammals relatively larger than that of other vertebrates, the cerebrum is convoluted. Convolutions (gyri and sulci) increase the surface area where cell bodies of neurons are located.

5. *Three auditory ossicles* — The auditory ossicles (bones) function as levers in amplifying sound waves passing through the middle-ear chamber.

6. *Double occipital condyle* — Also found in amphibians, the double occipital condyle provides support of the head on the vertebral column and permits extensive flexibility.

7. *Single dentary bone* — The lower jaw, or mandible, consists of a single bone that articulates at the skull at the temporomandibular joint.

8. *Seven cervical vertebrae* — Although a few species of mammals do not have seven cervical (neck) vertebrae, this is generally considered a diagnostic mammalian trait.

9. *Heterodont dentition* — Teeth that differ in structure and are adapted to handle food in different ways. The distinct teeth include incisors, canines, premolars, and molars.

10. *Pinnea (fleshy outer ears)* and *movable eyelids* — Each pinna funnels sound waves to the auditory tube; the movable eyelids protect the anterior surface of the eyeball and keep it from drying out.

11. *Muscular diaphragm* — Separating the thoracic and abdominal cavities, the muscular diaphragm is the principal muscle of inspiration. An effective respiratory system is essential for maintaining homeothermia.

12. *Nonnucleated, biconcave red blood cells* — The red blood cells, or erythrocytes, are produced in red bone marrow and transport oxygen to each of the living cells within the mammalian body.

13. *Placental attachment of fetus* — Mammals have internal fertilization, prenatal development is in a uterus with placental attachment (except in monotremes), and fetal membranes (amnion, chorion, allantois) protect and sustain the fetus.

The highly specialized mammalian brain has enabled these animals to learn and adapt in response to experiences, and to evolve intricate social structures within species populations.

Figure 17.1 Bottlenose dolphin, *Tursiops truncatus*, a highly social marine mammal.

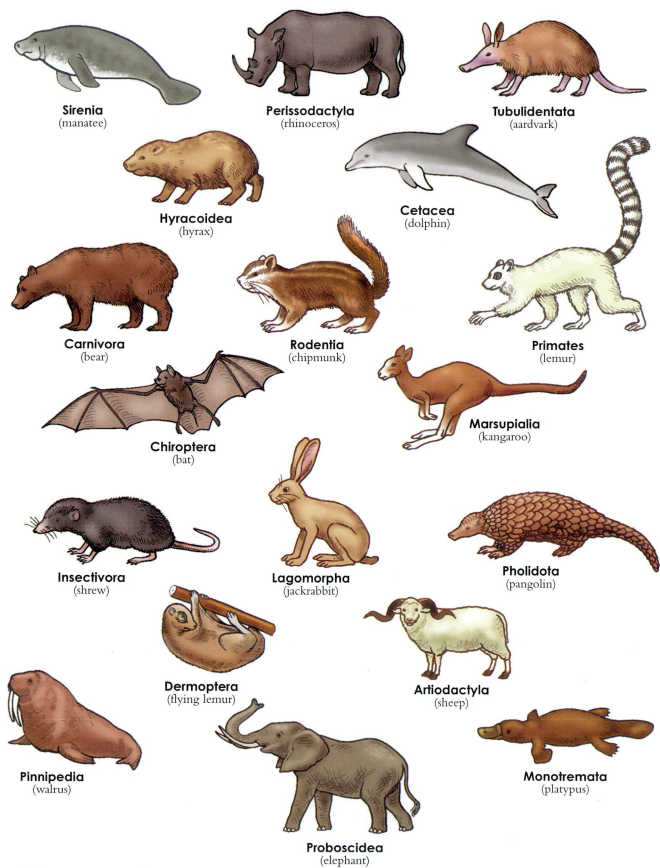

Sirenia
(manatee)

Perissodactyla
(rhinoceros)

Tubulidentata
(aardvark)

Hyracoidea
(hyrax)

Cetacea
(dolphin)

Carnivora
(bear)

Rodentia
(chipmunk)

Primates
(lemur)

Chiroptera
(bat)

Marsupialia
(kangaroo)

Insectivora
(shrew)

Lagomorpha
(jackrabbit)

Pholidota
(pangolin)

Dermoptera
(flying lemur)

Artiodactyla
(sheep)

Pinnipedia
(walrus)

Monotremata
(platypus)

Proboscidea
(elephant)

Figure 17.2 Representatives from the orders of mammals.

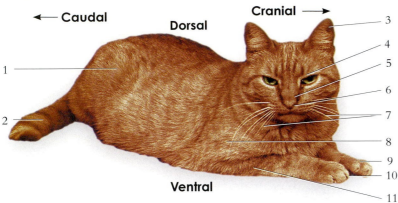

Figure 17.3 Directional terminology and superficial structures in a cat (quadrupedal vertebrate).

1. Thigh
2. Tail
3. Auricle (pinna)
4. Superior palpebra (superior eyelid)
5. Bridge of nose
6. Naris (nostril)
7. Vibrissae
8. Brachium
9. Manus (front foot)
10. Claw
11. Antebrachium

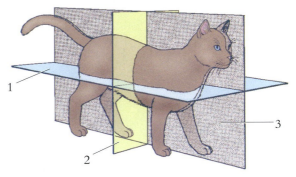

Figure 17.4 Planes of reference in a cat.

1. Coronal plane (frontal plane)
2. Transverse plane (cross-sectional plane)
3. Midsagittal plane (median plane)

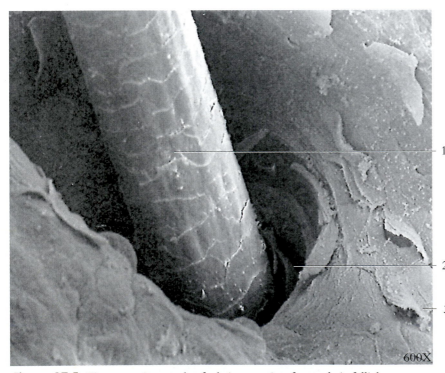

600X

Figure 17.5 Electron micrograph of a hair emerging from a hair follicle.
1. Shaft of hair (note the scale–like pattern)
2. Hair follicle
3. Epithelial cell from stratum corneum

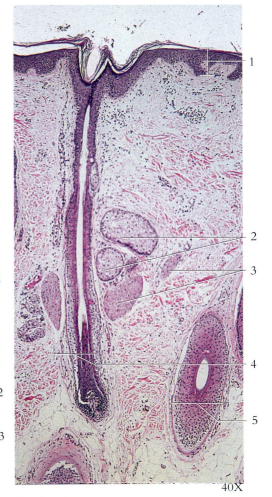

40X

Figure 17.6 Hair follicle.
1. Epidermis
2. Sebaceous glands
3. Arrector pili muscle
4. Hair follicle
5. Hair follicle (oblique cut)

Rat Dissection

The laboratory white rat is a captive-raised rodent that is commercially available for biological and medical experiments and research. White rats are also embalmed and available as dissection specimens in biology, vertebrate biology, and general zoology laboratories. Before the muscles and viscera of a rat can be studied, the specimen's skin has to be removed according to the following suggested guidelines.

1. Place the rat on a dissecting tray dorsal side up. Using a sharp scalpel, make a short, shallow incision through the skin across the neck. With your scissors, continue a dorsal midline incision to about two inches onto the tail. Sever the tail with bone shears and discard.

2. Make a shallow incision around the neck and down each foreleg to the paws. Continue a circular incision around each wrist. Beginning at the base of the tail, make incisions down each of the hind legs to the ankles. Make a circular cut around each ankle.

3. Carefully remove the skin, using your fingers or a blunt probe to separate the skin from underlying connective tissue. Where it is necessary to use a scalpel, keep the cutting edge directed toward the skin away from the muscle. If your specimen is a male, make an incision around the genitalia, leaving the skin intact. If your specimen is a female, the mammary glands will appear as longitudinal, glandular masses along the ventral side of the abdomen and thorax. They should be removed with the skin.

4. After the specimen is skinned, remove the excess fat and connective tissue to expose the underlying muscles. Make certain that the muscles are separated along their natural boundaries. If a transection of a muscle is necessary, isolate the muscle from its attached connective tissue and make a clean cut across the belly of the muscle, leaving the origin and insertion intact.

5. At the end of the laboratory period, wrap your specimen in muslin cloth and store it in a tight, heavy-duty plastic bag. Wet your specimen from time to time with a preservative solution (usually 2%–3% phenol). Caution needs to be taken when using a phenol wetting solution as it is caustic and poisonous if misused or used in a concentrated form.

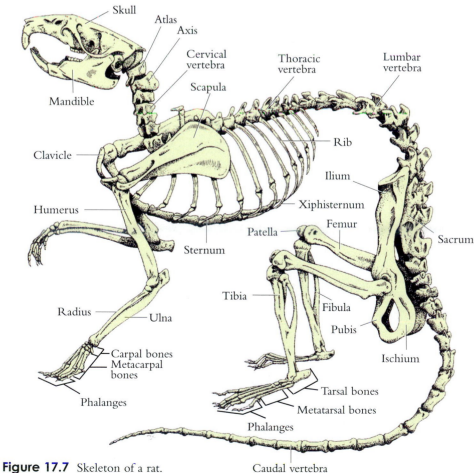

Figure 17.7 Skeleton of a rat.

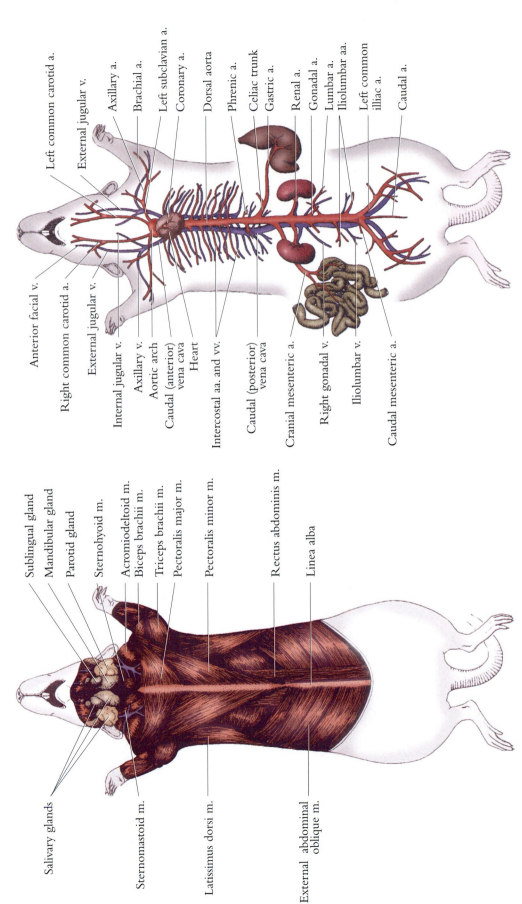

Figure 17.9 Circulatory system of a rat. The arteries are colored red (a. = artery, aa. = arteries; v. = vein, vv. = veins).

Left common carotid a.
External jugular v.
Axillary a.
Brachial a.
Left subclavian a.
Coronary a.
Dorsal aorta
Phrenic a.
Celiac trunk
Gastric a.
Renal a.
Gonadal a.
Lumbar a.
Iliolumbar aa.
Left common iliac a.
Caudal a.

Anterior facial v.
Right common carotid a.
External jugular v.
Internal jugular v.
Axillary v.
Aortic arch
Caudal (anterior) vena cava
Heart
Intercostal aa. and vv.
Caudal (posterior) vena cava
Cranial mesenteric a.
Right gonadal v.
Iliolumbar v.
Caudal mesenteric a.

Figure 17.8 Ventral view of the rat musculature (m. = muscle).

Sublingual gland
Mandibular gland
Parotid gland
Sternohyoid m.
Acromiodeltoid m.
Biceps brachii m.
Triceps brachii m.
Pectoralis major m.
Pectoralis minor m.
Rectus abdominis m.
Linea alba

Salivary glands
Sternomastoid m.
Latissimus dorsi m.
External abdominal oblique m.

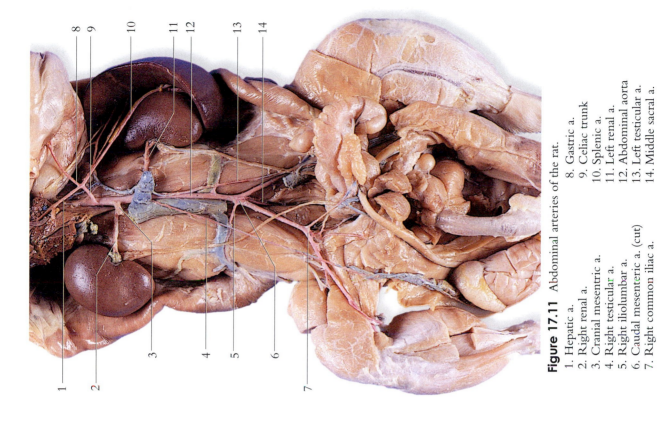

Figure 17.11 Abdominal arteries of the rat.

1. Hepatic a.
2. Right renal a.
3. Cranial mesenteric a.
4. Right testicular a.
5. Right iliolumbar a.
6. Caudal mesenteric a. (cut)
7. Right common iliac a.

8. Gastric a.
9. Celiac trunk
10. Splenic a.
11. Left renal a.
12. Abdominal aorta
13. Left testicular a.
14. Middle sacral a.

Figure 17.10 Ventral view of the rat viscera.

1. Trachea
2. Right lung
3. Right uterine horn of
   pregnant female
4. Jejunum
5. Cecum
6. Esophagus
7. Heart

8. Diaphragm (cut)
9. Liver
10. Stomach
11. Spleen
12. Ileum
13. Left uterine horn
    of pregnant female

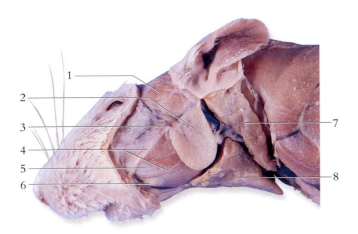

Figure 17.12 Head and neck region of the rat.

1. Temporalis muscle
2. Extraorbital lacrimal gland
3. Extraorbital lacrimal duct
4. Facial nerve
5. Masseter muscle
6. Parotid duct
7. Parotid gland
8. Mandibular gland

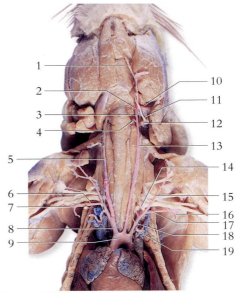

Figure 17.13 Arteries of the thoracic and neck regions of the rat.

1. Facial a.
2. Lingual a.
3. External carotid a.
4. Cranial thyroid a.
5. Common carotid a.
6. Axillary a.
7. Brachial a.
8. Brachiocephalic a.
9. Aortic arch
10. External maxillary a.
11. Internal carotid a.
12. Occipital a.
13. Common carotid a.
14. Vertebral a.
15. Cervical trunk
16. Lateral thoracic a.
17. Axillary a.
18. Subclavian a.
19. Internal thoracic a.

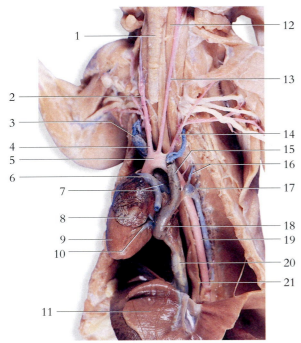

Figure 17.14 Reflected rat heart showing the major veins and arteries.

1. Trachea
2. Right common carotid a.
3. Right cranial vena cava
4. Brachiocephalic trunk
5. Aortic arch
6. Pulmonary trunk
7. Left and right pulmonary a.a.
8. Left auricle
9. Left ventricle
10. Coronary vein
11. Diaphragm
12. Esophagus
13. Left common carotid a.
14. Left subclavian a.
15. Left cranial vena cava
16. Intercostal a. and v.
17. Azygos v.
18. Coronary sinus
19. Aorta
20. Caudal vena cava
21. Esophagus

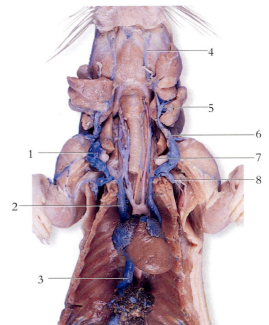

Figure 17.15 Veins of the thoracic and neck regions of the rat.

1. Cephalic v.
2. Cranial vena cava
3. Caudal vena cava
4. Linguofacial v.
5. Maxillary v.
6. External jugular v.
7. Internal jugular v.
8. Lateral thoracic v.

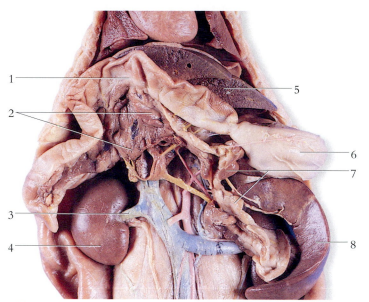

Figure 17.16 Abdominal viscera and vessels of the rat.

1. Duodenum
2. Biliary and duodenal parts of pancreas
3. Right renal v.
4. Right kidney
5. Liver (cut)
6. Stomach
7. Gastrosplenic part of pancreas
8. Spleen

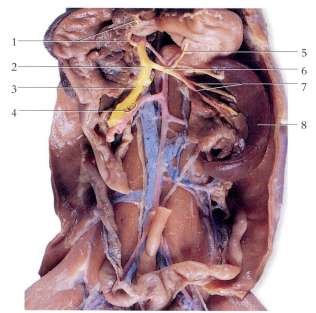

Figure 17.17 Branches of the hepatic portal system.

1. Cranial pancreaticoduodenal v.
2. Hepatic portal v.
3. Cranial mesenteric v.
4. Intestinal branches
5. Gastric v.
6. Gastrosplenic v.
7. Splenic branches
8. Spleen

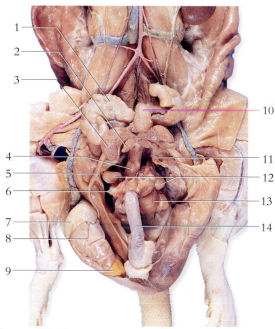

Figure 17.18 Urogenital system of the male rat.

1. Vesicular gland
2. Prostate (dorsolateral part)
3. Prostate (ventral part)
4. Urethra in the pelvic canal
5. Ductus (vas) deferens
6. Crus of penis (cut)
7. Head of epididymis
8. Testis
9. Tail of epididymis
10. Urinary bladder
11. Symphysis pubis (cut exposing pelvic canal)
12. Bulbourethral glands
13. Bulbocavernosus muscle
14. Penis

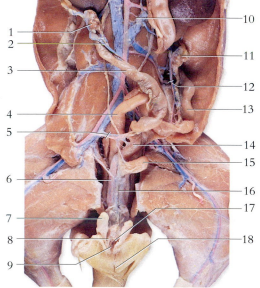

Figure 17.19 Urogenital system of the female rat.

1. Ovary
2. Uterine a. and v.
3. Uterine horn
4. Colon
5. Vesicular a. (umbilical a.)
6. Vagina
7. Preputial gland
8. Clitoris
9. Vaginal opening
10. Ovarian a. and v.
11. Ovary
12. Uterine a. and v.
13. Uterine horn
14. Uterine body
15. Urinary bladder
16. Urethra
17. Urethral opening
18. Anus

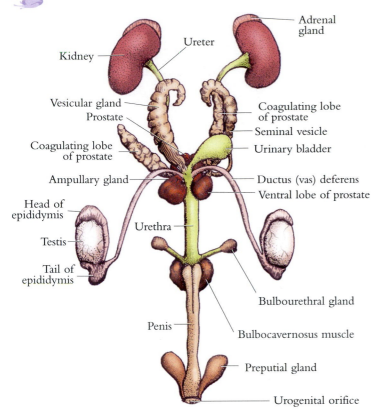

Figure 17.20 Urogenital organs of a male rat.

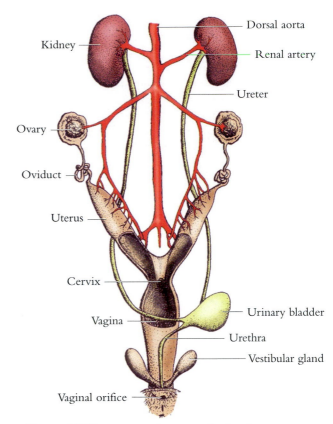

Figure 17.21 Urogenital organs of a female rat.

Fetal Pig Dissection

Fetal pigs are purchased from biological supply houses and are specially prepared for dissection. Excess embalming fluid should be drained from the packaged specimen prior to dissection.

Examine your specimen and identify the **umbilical cord.** Locate the two rows of **teats** that extend the length of the abdomen. Determine the sex of your specimen. A male has a **scrotal sac** in the pelvic region of the body between the hind legs and a **urogenital opening** just caudal to the umbilical cord. The **penis** can be palpated as a muscular tubular structure just underneath the skin along the midline proceeding caudally from the urogenital opening. A female has a small fleshy **genital papilla** projecting from the urogenital opening, which is located immediately ventral to the **anal opening.**

Before the muscles and viscera of a fetal pig can be studied, the specimen's skin has to be removed according to the following suggested guidelines.

1. Place your specimen on a dissecting tray ventral side up. Using a sharp scalpel, make a shallow incision through the skin extending from the chin caudally to the umbilical cord. Carefully continue your cut around one side of the umbilical cord. If your specimen is a male, make a diagonal cut from the umbilical cord to the scrotum. If a female, continue a midventral incision from the umbilical cord to the genital papilla. Make an incision around the genitalia and tail.

2. From the midventral incision, extend an incision down the medial surfaces of the forelegs to the hoofs and then do the same for the skin of the hindlegs. Make circular incisions around each of the hoofs. Following the ventral borders of the lower jaws, make extended cuts from the chin dorsolaterally to just below the ears.

3. Grasp the cut edge of the skin and carefully remove it from your specimen. If the skin is difficult to remove, grasp the cut edge of the skin with one hand and push on the muscle with the thumb of the other hand.

4. After the specimen is skinned, the muscles can be seen more easily if the moisture on them is sponged away with a paper towel. The muscles of a fetal pig are extremely delicate and as you proceed to dissect your specimen, make certain that you separate the muscles along their natural boundaries. When transection of a muscle is necessary, carefully isolate the muscle from its attached connective tissue and make a clean cut across the belly of the muscle, leaving the origin and insertion intact.

5. At the end of the laboratory period, wrap your specimen in muslin cloth and store it in a tight, heavy-duty plastic bag. Discard the skin that was removed from your specimen, and the plastic shipment bag. Wet your specimen from time to time with a preservative solution (usually 2-3% phenol). Caution is necessary when using a phenol wetting solution as it is caustic and poisonous if misused or used in a concentrated form.

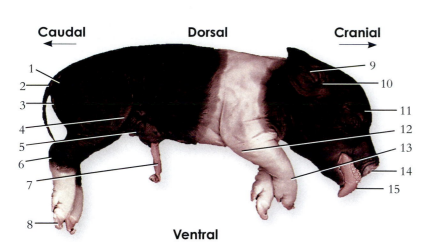

Caudal **Dorsal** **Cranial**

Ventral

Figure 17.22 Directional terminology and superficial structures in a fetal pig (quadrupedal vertebrate).

1. Anus	9. Auricle (pinna)
2. Tail	10. External auditory canal
3. Scrotum	11. Superior palpebra
4. Knee	(superior eyelid)
5. Penis	12. Elbow
6. Ankle	13. Wrist
7. Umbilical cord	14. Naris (nostril)
8. Hoof	15. Tongue

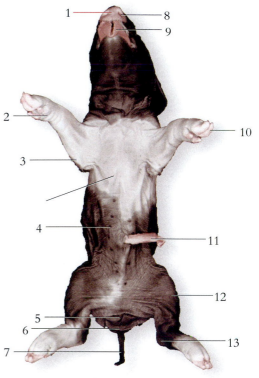

Figure 17.23 Ventral view of the surface anatomy of the fetal pig.

1. Nose	6. Scrotum	11. Umbilical cord
2. Wrist	7. Tail	12. Knee
3. Elbow	8. Nostril	13. Ankle
4. Teats	9. Tongue	
5. Anus	10. Digit	

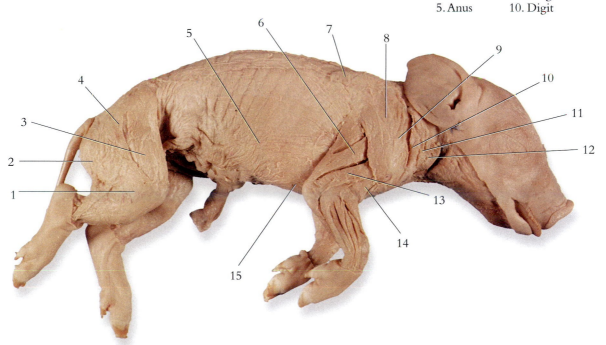

Figure 17.24 Lateral view of superficial musculature of the fetal pig.

1. Biceps femoris m.	6. Triceps brachii m.	10. Cleidooccipitalis m.	14. Brachialis m.
2. Semitendinosus m.	(long head)	11. Cleidomastoid m.	15. Pectoralis profundus m.
3. Tensor fasciae latae m.	7. Trapezius m.	12. Sternocephalicus m.	
4. Gluteus medius m.	8. Deltoid m.	13. Triceps brachii m.	
5. External abdominal oblique m.	9. Supraspinatus m.	(lateral head)	

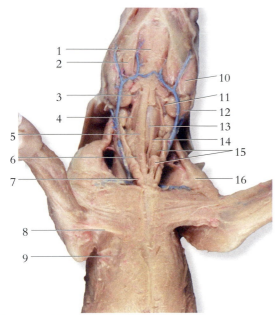

Figure 17.25 Ventral view of superficial muscles of neck and upper torso.

1. Mylohyoid m.
2. Digastric m.
3. Stylohyoid m.
4. Omohyoid m.
5. Sternohyoid m.
6. Thymus
7. Sternomastoid m.
8. Pectoralis superficialis m.
9. Pectoralis profundus m.
10. Masseter m.
11. Thyrohyoid m.
12. Mandibular gland
13. Larynx
14. Sternothyroid m.
15. Mandibular lymph nodes
16. Brachiocephalic m.

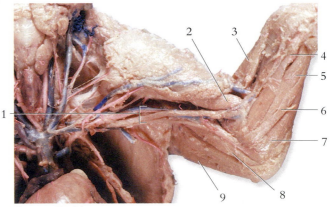

Figure 17.26 Superficial medial mucles of the forelimb.

1. Axillary a. and v., brachial plexus
2. Biceps brachii m.
3. Extensor carpi radialis m.
4. Flexor carpi radialis m.
5. Flexor digitorum profundus m.
6. Flexor digitorum superficialis m.
7. Flexor carpi ulnaris m.
8. Triceps brachii m. (lateral head)
9. Triceps brachii m. (long head)

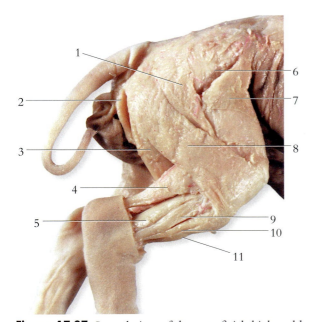

Figure 17.27 Lateral view of the superficial thigh and leg.

1. Gluteus superficialis m.
2. Semitendinosus m.
3. Semimembranosus m.
4. Gastrocnemius m.
5. Extensor digitorum quarti and quinti mm.
6. Gluteus medius m.
7. Tensor fasciae latae m.
8. Biceps femoris m.
9. Peroneus longus m.
10. Peroneus tertius m.
11. Tibialis anterior m.

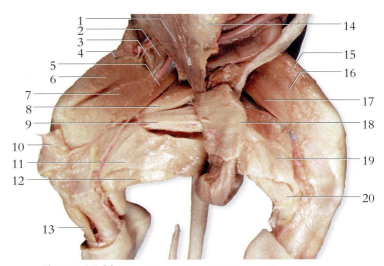

Figure 17.28 Medial muscles of thigh and leg.

1. External abdominal oblique m.
2. Psoas major m.
3. Iliacus m.
4. Tensor fasciae latae m.
5. Sartorius m.
6. Rectus femoris m.
7. Vastus medialis m.
8. Pectineus m.
9. Adductor m.
10. Aponeurosis of gracilis m. (cut)
11. Semimembranosus m.
12. Semitendinosus m.
13. Tibialis anterior m.
14. Linea alba
15. Rectus femoris m.
16. Vastus medialis m.
17. Sartorius m.
18. Gracilis m. (cut)
19. Gracilis m.
20. Semitendinosus m.

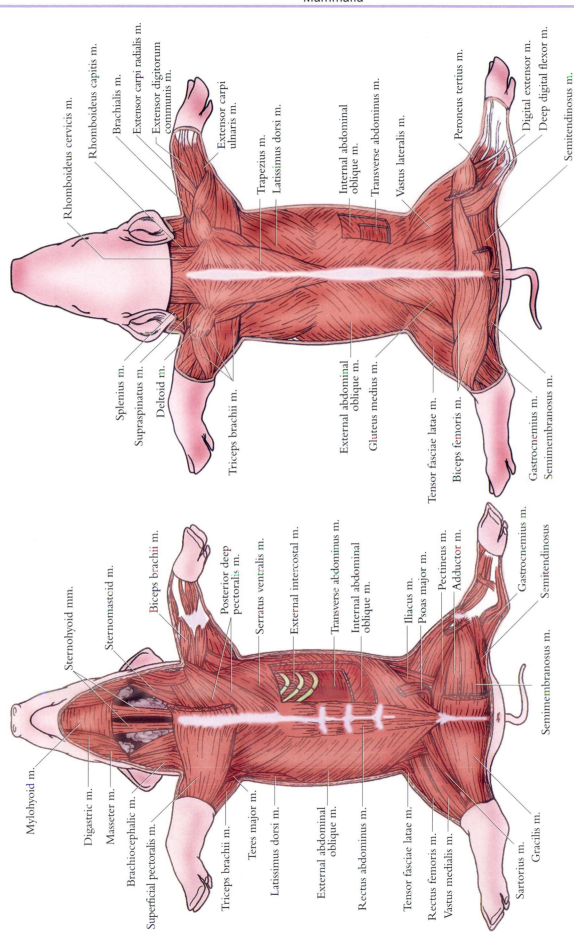

Rhomboideus cervicis m.

Rhomboideus capitis m.

Brachialis m.

Extensor carpi radialis m.

Extensor digitorum communis m.

Extensor carpi ulnaris m.

Trapezius m.

Latissimus dorsi m.

Internal abdominal oblique m.

Transverse abdominus m.

Vastus lateralis m.

Peroneus tertius m.

Digital extensor m.

Deep digital flexor m.

Semitendinosus m.

Splenius m.

Supraspinatus m.

Deltoid m.

Triceps brachii m.

External abdominal oblique m.

Gluteus medius m.

Tensor fasciae latae m.

Biceps femoris m.

Gastrocnemius m.

Semimembranosus m.

Figure 17.30 Dorsal view of the muscles of the fetal pig.

Biceps brachii m.

Posterior deep pectoralis m.

Serratus ventralis m.

External intercostal m.

Transverse abdominus m.

Internal abdominal oblique m.

Iliacus m.

Psoas major m.

Pectineus m.

Adductor m.

Gastrocnemius m.

Semitendinosus

Sternohyoid mm.

Sternomastoid m.

Semimembranosus m.

Mylohyoid m.

Digastric m.

Masseter m.

Brachiocephalic m.

Superficial pectoralis m.

Triceps brachii m.

Teres major m.

Latissimus dorsi m.

External abdominal oblique m.

Rectus abdominus m.

Tensor fasciae latae m.

Rectus femoris m.

Vastus medialis m.

Sartorius m.

Gracilis m.

Figure 17.29 Ventral view of the muscles of the fetal pig.

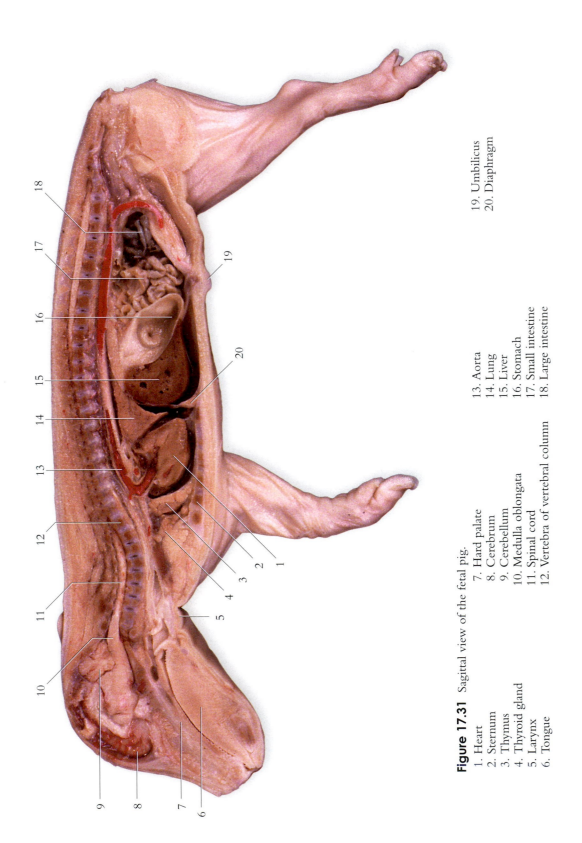

Figure 17.31 Sagittal view of the fetal pig.

1. Heart
2. Sternum
3. Thymus
4. Thyroid gland
5. Larynx
6. Tongue

7. Hard palate
8. Cerebrum
9. Cerebellum
10. Medulla oblongata
11. Spinal cord
12. Vertebra of vertebral column

13. Aorta
14. Lung
15. Liver
16. Stomach
17. Small intestine
18. Large intestine

19. Umbilicus
20. Diaphragm

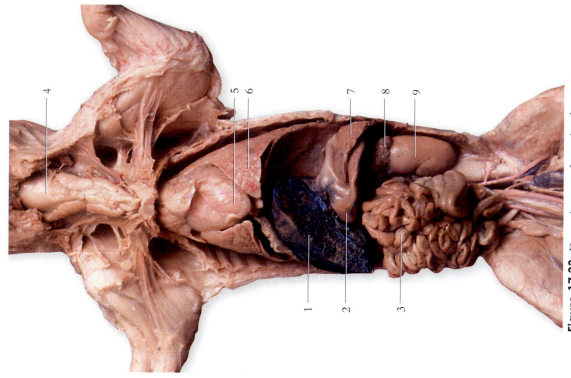

Figure 17.33 Deep viscera and associated structures.

1. Liver (cut)
2. Stomach
3. Small intestine
4. Larynx
5. Heart
6. Lung
7. Spleen
8. Adrenal gland
9. Kidney

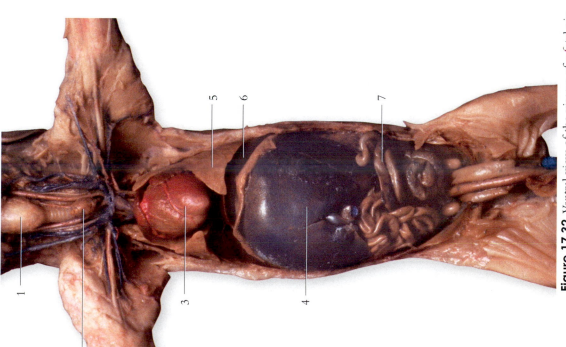

Figure 17.32 Ventral view of the viscera of a fetal pig.

1. Larynx
2. Thyroid gland
3. Heart
4. Liver
5. Lung
6. Diaphragm
7. Small intestine

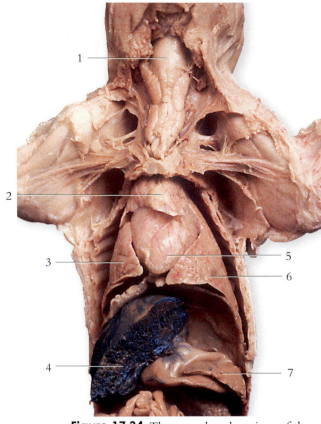

Figure 17.34 Thorax and neck regions of the fetal pig.

1. Larynx	5. Heart
2. Thymus	6. Lung
3. Lung	7. Spleen (cut)
4. Liver (cut)	

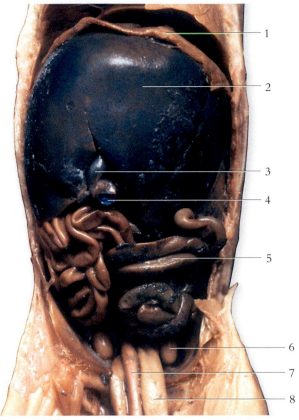

Figure 17.35 Ventral view of the abdominal cavity of a fetal pig.

1. Diaphragm	5. Small intestine
2. Liver	6. Undescended testis
3. Gallbladder	7. Umbilical artery
4. Umbilical vein	8. Urinary bladder

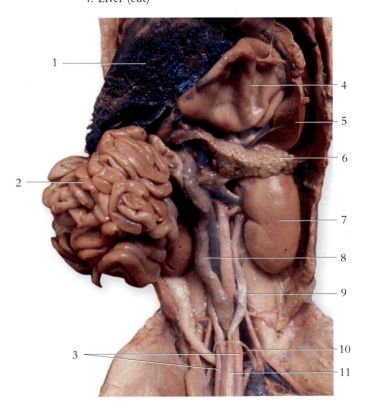

Figure 17.36 Abdominal organs of the fetal pig.

1. Liver (cut)	7. Kidney
2. Small intestine	8. Large intestine
3. Umbilical a.a.	9. Ureter
4. Stomach (reflected)	10. Ductus (vas) deferens
5. Spleen	11. Urinary bladder
6. Pancreas	

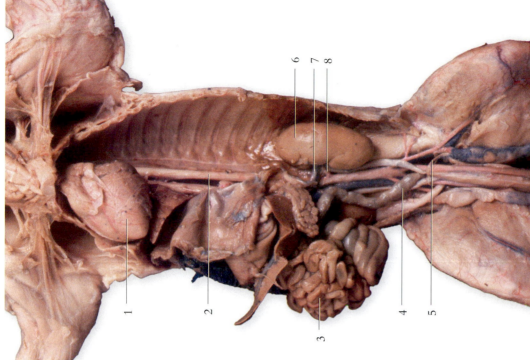

Figure 17.38 Blood supply to the abdomen and lower extremities of the fetal pig.

1. Heart
2. Thoracic aorta
3. Small intestine
4. Colon
5. Ductus deferens
6. Kidney
7. Renal a.
8. Renal v.

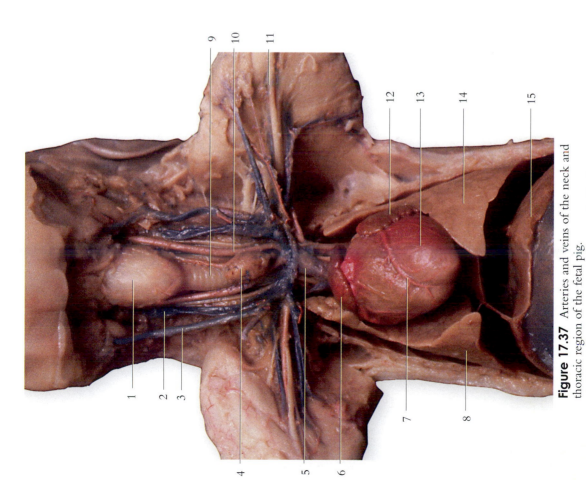

Figure 17.37 Arteries and veins of the neck and thoracic region of the fetal pig.

1. Larynx
2. Internal jugular v.
3. External jugular v.
4. Thyroid gland
5. Cranial (superior) vena cava
6. Right auricle
7. Coronary vessels
8. Right lung
9. Trachea
10. Left common carotid a.
11. Axillary a.
12. Left auricle
13. Left ventricle
14. Left lung
15. Diaphragm

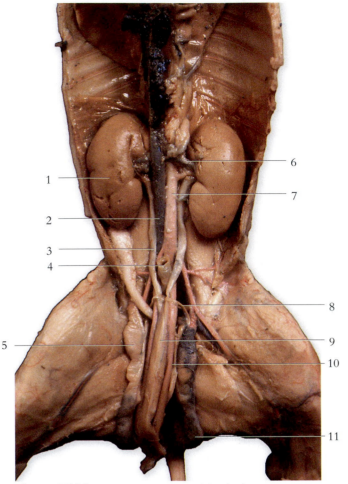

Figure 17.39 Urogenital system of the fetal pig.

1. Kidney
2. Caudal (inferior) vena cava
3. Ureter
4. Rectum (cut)
5. Partially dissected testis
6. Renal v.
7. Descending aorta
8. Ductus deferens
9. Urinary bladder
10. Umbilical a.
11. Epididymis

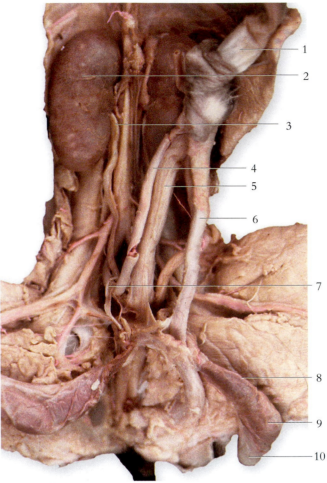

Figure 17.40 Urogenital system of the fetal pig.

1. Umbilical cord
2. Right kidney
3. Ureter
4. Umbilical a.
5. Urinary bladder
6. Penis
7. Vas (ductus) deferens
8. Spermatic cord
9. Right testis
10. Epididymis

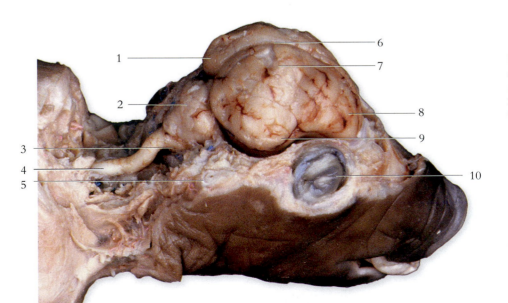

Figure 17.41 General structures of the fetal pig brain. Because the cerebrum is less defined in pigs, the regions are not known as lobes as they are in humans.

1. Occipital region of cerebrum
2. Cerebellum
3. Medulla oblongata
4. Spinal cord
5. External acoustic meatus
6. Longitudinal fissure
7. Parietal region of cerebrum
8. Frontal region of cerebrum
9. Temporal region of cerebrum
10. Eye

Rabbit Dissection

Embalmed rabbits are are often used in learning basic mammalian anatomy. Before the muscles and viscera be studied, the specimen's skin has to be removed according to the following suggested guidelines.

1. Place the rabbit on a dissecting tray dorsal side up. Using a sharp scalpel, make a short, shallow incision through the skin across the nape of the neck. With your scissors, continue a dorsal midline incision forward over the skull and down the back to the tail.

2. Make a shallow incision around the neck and down each foreleg to the paws. Continue a circular incision around each wrist. Beginning at the base of the tail, make incisions down each of the hind legs to the ankles. Make a circular cut around each ankle, and around the tail.

3. Carefully remove the skin, using your fingers or a blunt probe to separate the skin from underlying connective tissue. Where it is necessary to use a scalpel, keep the cutting edge directed toward the skin away from the muscle. If your specimen is a male, make an incision around the genitalia, leaving the skin intact. If your specimen is a female, the mammary glands will appear as longitudinal, glandular masses along the ventral side of the abdomen and thorax. They should be removed with the skin.

4. After the specimen is skinned, remove the excess fat and connective tissue to expose the underlying muscles. Make certain that the muscles are separated along their natural boundaries. If a transection of a muscle is necessary, isolate the muscle from its attached connective tissue and make a clean cut across the belly of the muscle, leaving the origin and insertion intact.

5. At the end of the laboratory period, wrap your specimen in muslin cloth and store it in a tight, heavy-duty plastic bag. Wet your specimen from time to time with a preservative solution (usually 2%-3% phenol). Caution needs to be taken when using a phenol wetting solution as it is caustic and poisonous if misused or used in a concentrated form.

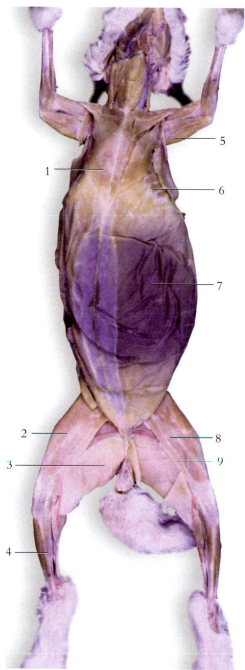

Figure 17.42 Ventral view of the superficial musculature of the rabbit. (m. = muscle)

1. Pectoralis major m.
2. Tensor fasciae latae m.
3. Gracilis m.
4. Tibialis anterior m.
5. Triceps brachii m. (long head)
6. Serratus ventralis m.
7. External abdominal oblique m.
8. Vastus medialis m.
9. Sartorius m.

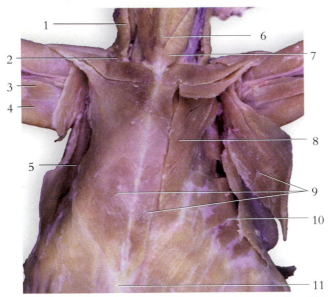

Figure 17.43 Neck and thoracic musculature of the rabbit.

1. Basioclavicularis m.
2. Pectoralis tenius m.
3. Epitrochlearis m.
4. Triceps brachii m. (long head)
5. Teres major m.
6. Sternohyoid m.
7. Sternomastoid m.
8. Pectoralis profundus m.
9. Pectoralis major m.
10. Serratus ventralis m.
11. Linea alba

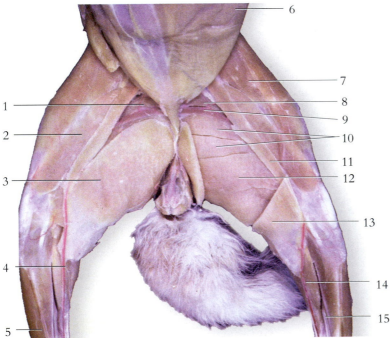

Figure 17.44 Superficial musculature of the abdomen and hind limbs of the rabbit.

1. Pectineus m.
2. Vastus medialis m.
3. Gracilis m.
4. Gastrocnemius m.
5. Tibialis anterior m.
6. External abdominal oblique m.
7. Rectus femoris m.
8. Pectineus m.
9. Adductor brevis m.
10. Adductor longus m.
11. Sartorius m.
12. Adductor magnus m.
13. Gracilis m. (cut)
14. Soleus m.
15. Flexor digitorum longus m.

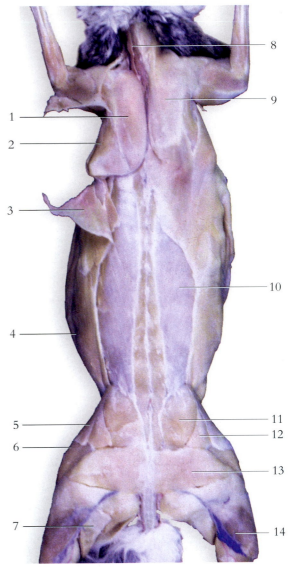

Figure 17.45 Dorsal view of the superficial musculature of the rabbit.

1. Supraspinatus m.
2. Infraspinatus m.
3. Spinotrapezius m. (cut)
4. External abdominal oblique m.
5. Tensor fasciae latae m.
6. Vastus lateralis m.
7. Semimembranosus proprius m.
8. Rhomboideus capitis m.
9. Acromiotrapezius m.
10. Longissimus dorsi m.
11. Gluteus medius m.
12. Rectus femoris m.
13. Biceps femoris m. (cranial portion)
14. Biceps femoris m. (caudal portion)

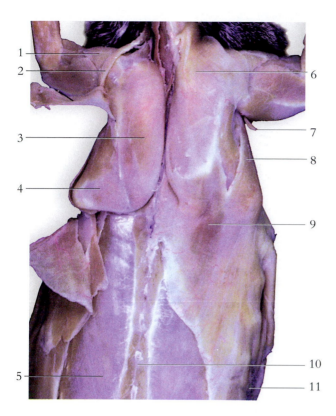

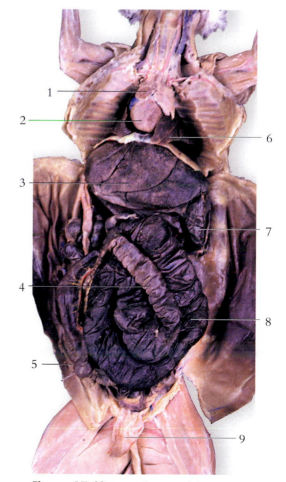

Figure 17.46 Dorsal view of the musculature of the shoulder thoracic and abdominal regions of the rabbit.

1. Cleidodeltoid m.
2. Levator scapulae ventralis m.
3. Supraspinatus m.
4. Infraspinatus m.
5. Longissimus dorsi m.
6. Acromiotrapezius m.
7. Cutaneous maximus m. (cut)
8. Latissimus dorsi m.
9. Spinotrapezius m.
10. Multifious m.
11. External abdominal oblique m.

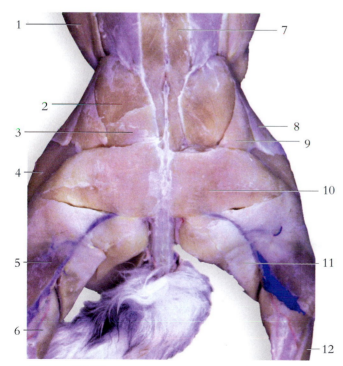

Figure 17.47 Dorsal view of the musculature of the hip and hind limbs of the rabbit.

1. Longissimus dorsi m.
2. Gluteus medius m.
3. Gluteus maximus m.
4. Vastus lateralis m.
5. Biceps femoris m. (caudal portion)
6. Gastrocnemius m.
7. Multifidus m.
8. Tensor fasciae latae m.
9. Rectus femoris m.
10. Biceps femoris m. (cranial portion)
11. Semimembranosus m.
12. Tibialis anterior m.

Figure 17.48 Ventral view of the rabbit viscera.

1. Thymus gland
2. Heart
3. Liver
4. Ileum of small intestine
5. Cecum
6. Left lung
7. Stomach
8. Colon
9. Urinary bladder

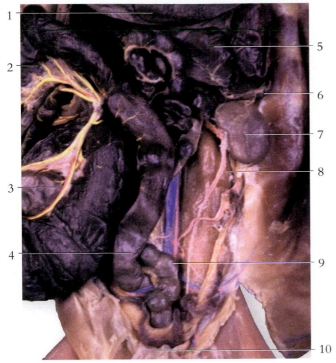

Figure 17.49 Ventral view of the abdominal viscera of the rabbit.

1. Liver	4. Appendix	8. Ureter
2. Cecum	5. Stomach	9. Descending
3. Sacculus	6. Spleen	colon
rotundus	7. Kidney	10. Urinary bladder

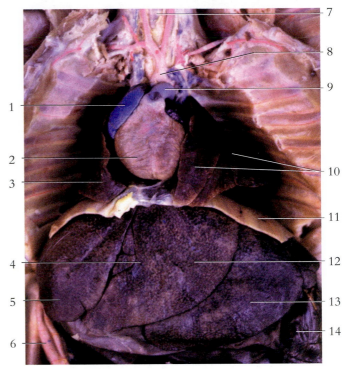

Figure 17.50 Ventral view of the viscera of the thoracic and upper abdominal cavities of the rabbit.

1. Right atrium of heart	8. Aortic arch
2. Right ventricle of heart	9. Pulmonary artery
3. Right lung	10. Left lung
4. Quadrate lobe of liver	11. Diaphragm
5. Right lobe of liver	12. Left median lobe of liver
6. Duodenum	13. Left lateral lobe of liver
7. Trachea	14. Stomach

Cat Dissection

Embalmed cats purchased from biological supply houses are excellent specimens for dissecting and learning basic mammalian anatomy. Before the muscles and viscera of a cat can be studied, the specimen's skin has to be removed according to the following suggested guidelines.

1. Place the cat on a dissecting tray dorsal side up. Using a sharp scalpel, make a short, shallow incision through the skin across the nape of the neck. With your scissors, continue a dorsal midline incision forward over the skull and down the back to about two inches onto the tail. Sever the tail with bone shears or a saw and discard.

2. Make a shallow incision around the neck and down each foreleg to the paws. Continue a circular incision around each wrist. Beginning at the base of the tail, make incisions down each of the hind legs to the ankles. Make a circular cut around each ankle.

3. Carefully remove the skin, using your fingers or a blunt probe to separate the skin from underlying connective tissue. Where it is necessary to use a scalpel, keep the

cutting edge directed toward the skin away from the muscle. If your specimen is a male, make an incision around the genitalia, leaving the skin intact. If your specimen is a female, the mammary glands will appear as longitudinal, glandular masses along the ventral side of the abdomen and thorax. They may be removed with the skin.

4. After the specimen is skinned, remove the excess fat and connective tissue to expose the underlying muscles. Make certain that the muscles are separated along their natural boundaries. If a transection of a muscle is necessary, isolate the muscle from its attached connective tissue and make a clean cut across the belly of the muscle, leaving the origin and insertion intact.

5. At the end of the laboratory period, wrap your specimen in muslin cloth and store it in a tight, heavy-duty plastic bag. Wet your specimen from time to time with a preservative solution (usually 2%–3% phenol). Caution needs to be taken when using a phenol wetting solution as it is caustic and poisonous if misused or used in a concentrated form.

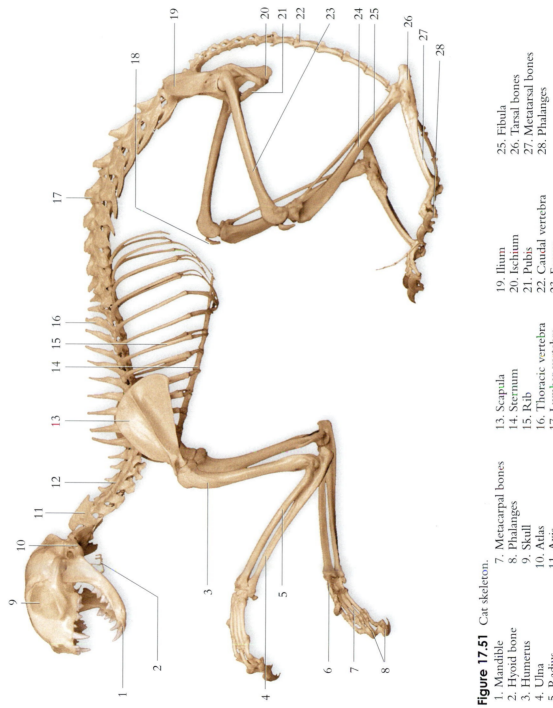

Figure 17.51 Cat skeleton.

1. Mandible	13. Scapula	25. Fibula
2. Hyoid bone	14. Sternum	26. Tarsal bones
3. Humerus	15. Rib	27. Metatarsal bones
4. Ulna	16. Thoracic vertebra	28. Phalanges
5. Radius	17. Lumbar vertebra	
6. Carpal bones	18. Patella	
7. Metacarpal bones	19. Ilium	
8. Phalanges	20. Ischium	
9. Skull	21. Pubis	
10. Atlas	22. Caudal vertebra	
11. Axis	23. Femur	
12. Cervical vertebra	24. Tibia	

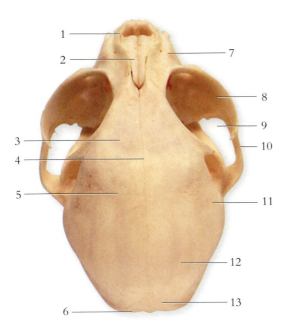

Figure 17.52 Dorsal view of a cat skull.
1. Premaxilla
2. Nasal bone
3. Frontal bone
4. Sagittal suture
5. Coronal suture
6. Nuchal crest
7. Maxilla
8. Zygomatic (malar) bone
9. Orbit
10. Zygomatic arch
11. Temporal bone
12. Parietal bone
13. Interparietal bone

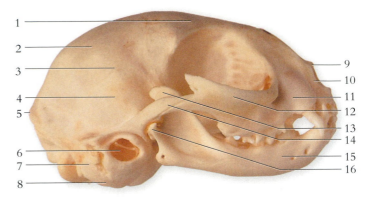

Figure 17.53 Lateral view of a cat skull.
1. Frontal bone
2. Parietal bone
3. Squamosal suture
4. Temporal bone
5. Nuchal crest
6. External acoustic meatus
7. Mastoid process
8. Tympanic bulla
9. Nasal bone
10. Premaxilla bone
11. Maxilla
12. Zygomatic (malar) bone
13. Coronoid process of mandible
14. Zygomatic arch
15. Mandible
16. Condylar process of mandible

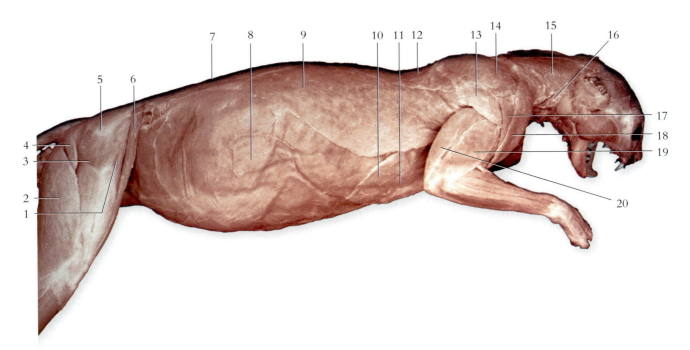

Figure 17.54 Lateral view of the superficial muscles of the cat.
1. Tensor fasciae latae m.
2. Biceps femoris m.
3. Gluteus maximus m.
4. Caudofemoralis m.
5. Gluteus medius m.
6. Sartorius m.
7. Lumbodorsal fascia
8. External abdominal oblique m.
9. Latissimus dorsi m.
10. Xiphihumeralis m.
11. Pectoralis minor m.
12. Spinotrapezius m.
13. Spinodeltoid m.
14. Acromiotrapezius m.
15. Clavotrapezius m.
16. Sternomastoid m.
17. Acromiodeltoid m.
18. Clavobrachialis m.
19. Lateral head of triceps brachii m.
20. Long head of triceps brachii m.

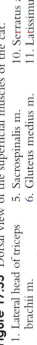

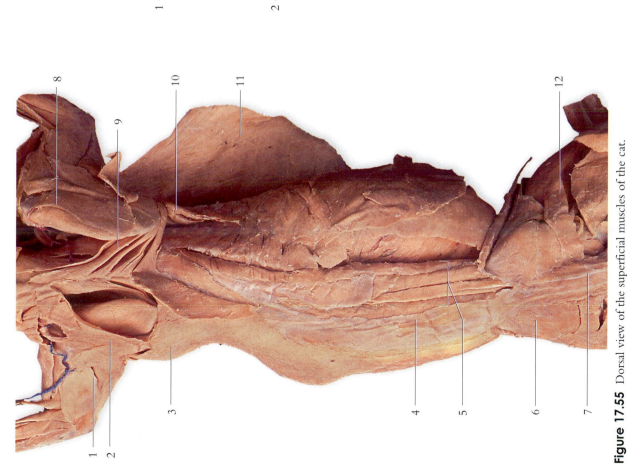

Figure 17.56 Dorsal view of the cat neck and thorax.

1. Temporalis m.
2. Clavotrapezius m.
3. Acromiotrapezius m.
4. Latissimus dorsi m.
5. Supraspinatus m.
6. Rhomboideus m.
7. Serratus anterior m.

Figure 17.55 Dorsal view of the superficial muscles of the cat.

1. Lateral head of triceps brachii m.
2. Acromiotrapezius m.
3. Latissimus dorsi m.
4. Lumbodorsal fascia
5. Sacrospinalis m.
6. Gluteus medius m.
7. Caudal m.
8. Supraspinatus m.
9. Rhomboideus m.
10. Serratus anterior m.
11. Latissimus dorsi m.
12. Caudofemoralis m.

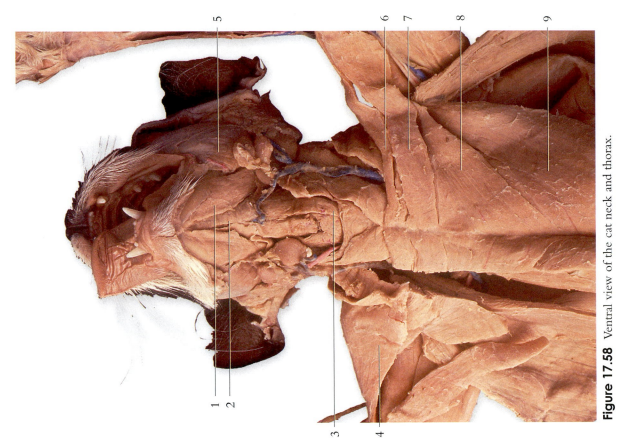

Figure 17.58 Ventral view of the cat neck and thorax.

1. Digastric m.
2. Mylohyoid m.
3. Sternomastoid m.
4. Clavotrapezius m.
5. Masseter m.
6. Clavobrachialis m.
7. Pectoantebrachialis m.
8. Pectoralis major m.
9. Pectoralis minor m.

Figure 17.57 Superficial view of the cat ventral trunk.

1. Mammary glands
2. Nipples
3. External abdominal oblique m.

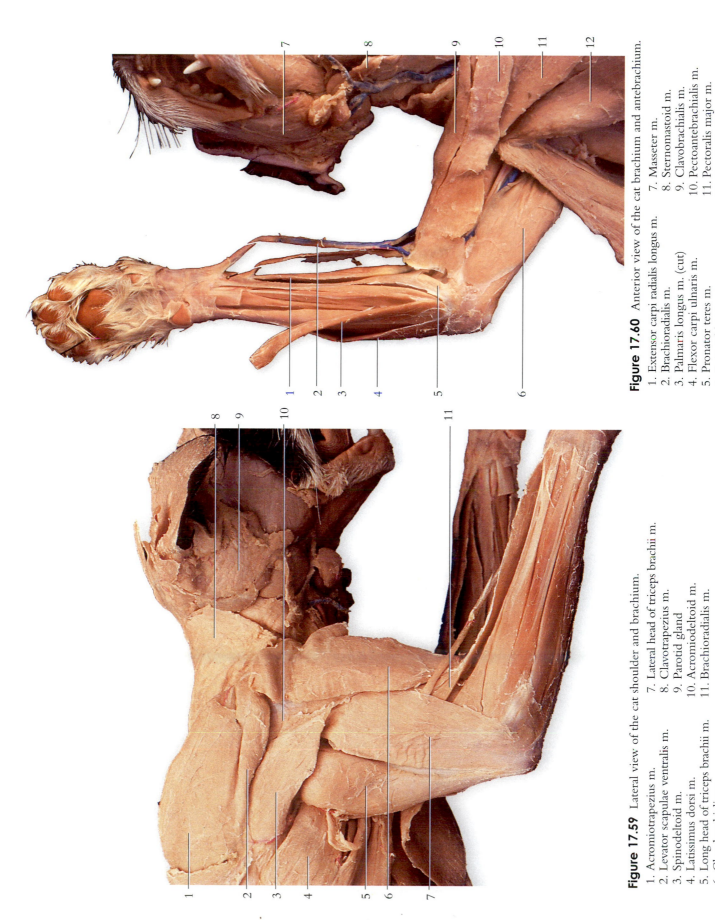

Figure 17.60 Anterior view of the cat brachium and antebrachium.

1. Extensor carpi radialis longus m.
2. Brachioradialis m.
3. Palmaris longus m. (cut)
4. Flexor carpi ulnaris m.
5. Pronator teres m.
6. Epitrochlearis
7. Masseter m.
8. Sternomastoid m.
9. Clavobrachialis m.
10. Pectoantebrachialis m.
11. Pectoralis major m.
12. Pectoralis minor m.

Figure 17.59 Lateral view of the cat shoulder and brachium.

1. Acromiotrapezius m.
2. Levator scapulae ventralis m.
3. Spinodeltoid m.
4. Latissimus dorsi m.
5. Long head of triceps brachii m.
6. Clavobrachialis m.
7. Lateral head of triceps brachii m.
8. Clavotrapezius m.
9. Parotid gland
10. Acromiodeltoid m.
11. Brachioradialis m.

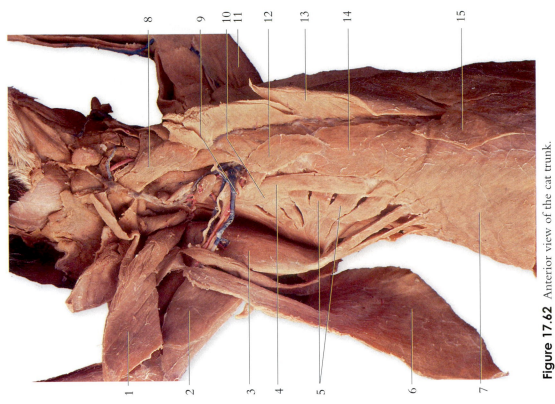

Figure 17.62 Anterior view of the cat trunk.

1. Pectoralis minor (cut)
2. Epitrochlearis m.
3. Subscapularis m.
4. Scalenus medius m.
5. Serratus anterior m.
6. Latissimus dorsi m. (cut)
7. External abdominal oblique m.
8. Sternomastoid m.
9. Scalenus anterior m.
10. Scalenus posterior m.
11. Epitrochlearis m.
12. Transverse costarum m.
13. Pectoralis minor m. (cut)
14. Rectus abdominis m.
15. Xiphihumeralis m. (cut)

Figure 17.61 Posterior view of the cat brachium and antebrachium.

1. Clavobrachialis m.
2. Acromiotrapezius m.
3. Brachioradialis m.
4. Extensor digitorum lateralis m.
5. Extensor digitorum communis m.
6. Extensor carpi radialis longus m.
7. Lateral head of triceps brachii m.
8. Long head of triceps brachii m.
9. Spinodeltoid m.
10. Latissimus dorsi m.

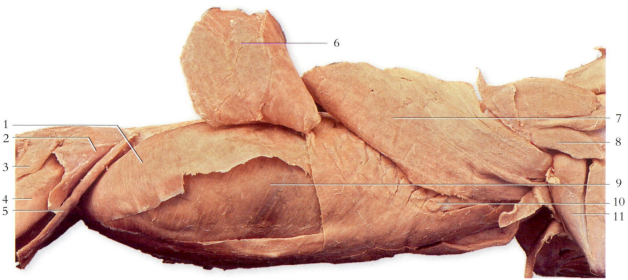

Figure 17.63 Lateral view of the cat trunk.

1. Internal abdominal oblique m.
2. Tensor fascia latae
3. Caudofemoris m.
4. Vastus lateralis m.
5. Sartorius m.
6. External abdominal oblique m.
7. Latissimus dorsi m.
8. Spinodeltoid m.
9. Transverse abdominis m.
10. Serratus anterior m.
11. Long head of triceps brachii m.

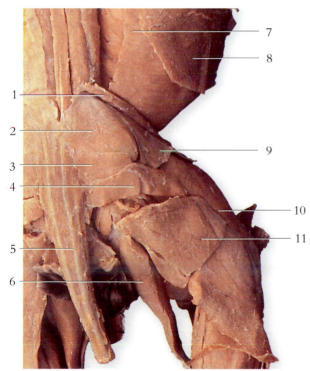

Figure 17.64 Lateral view of the cat superficial thigh.

1. Sartorius m.
2. Gluteus medius m.
3. Gluteus maximus m.
4. Caudofemoris m.
5. Caudal m.
6. Semitendinosus m.
7. Internal abdominal oblique m.
8. External abdominal oblique m.
9. Tensor fascia latae (cut)
10. Vastus lateralis m.
11. Biceps femoris m.

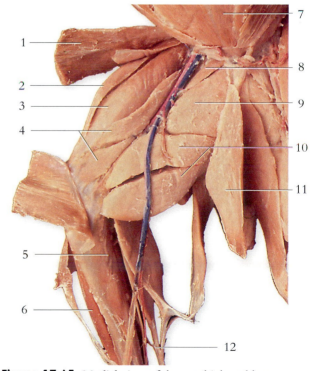

Figure 17.65 Medial view of the cat thigh and leg.

1. Sartorius m.
2. Vastus lateralis m.
3. Rectus femoris m.
4. Vastus medialis m.
5. Flexor digitorum longus m.
6. Tibialis anterior m.
7. Rectus abdominus m.
8. Adductor longus m.
9. Adductor femoris m.
10. Semimembranosus m.
11. Gracilis m. (cut)
12. Tendo calcaneus (Achilles tendon)

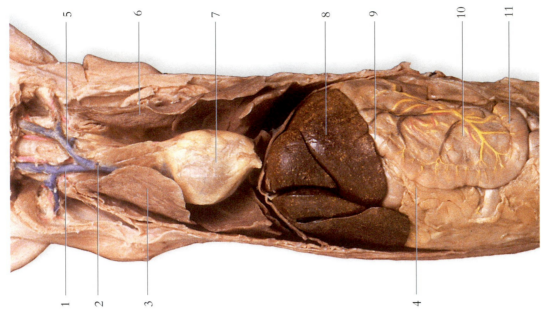

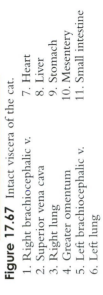

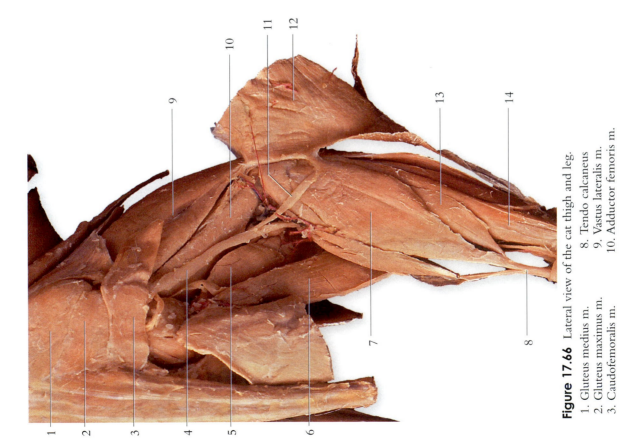

Figure 17.67 Intact viscera of the cat.

1. Right brachiocephalic v.
2. Superior vena cava
3. Right lung
4. Greater omentum
5. Left brachiocephalic v.
6. Left lung
7. Heart
8. Liver
9. Stomach
10. Mesentery
11. Small intestine

Figure 17.66 Lateral view of the cat thigh and leg.

1. Gluteus medius m.
2. Gluteus maximus m.
3. Caudofemoralis m.
4. Sciatic nerve
5. Semimembranosus m.
6. Semitendinosus m.
7. Gastrocnemius m.
8. Tendo calcaneus
9. Vastus lateralis m.
10. Adductor femoris m.
11. Tenuissimus m.
12. Biceps femoris m. (cut)
13. Soleus m.
14. Peroneal m.

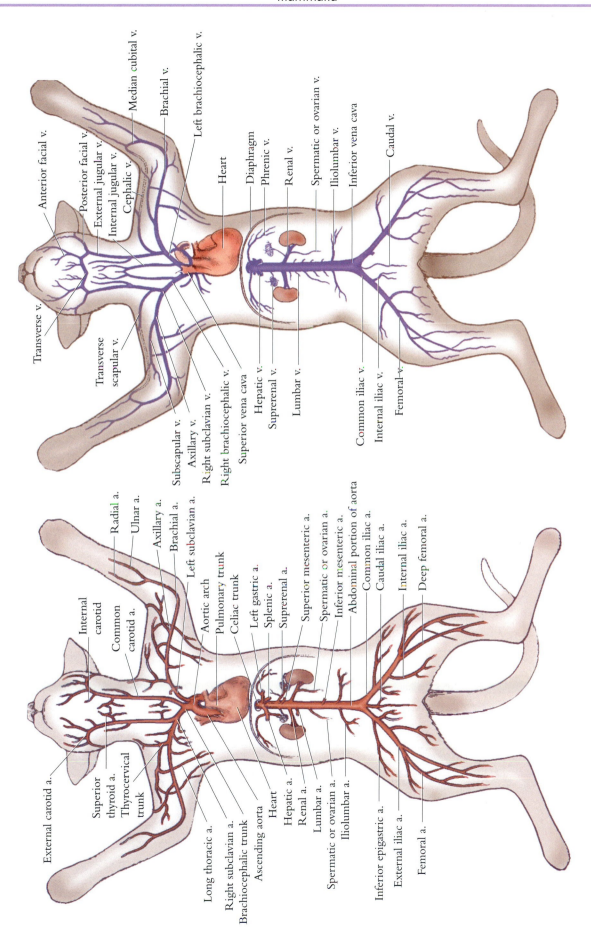

Figure 17.69 Principal veins of the cat, ventral view. (v=vein)

Figure 17.68 Principal arteries of the cat, ventral view. (a=artery)

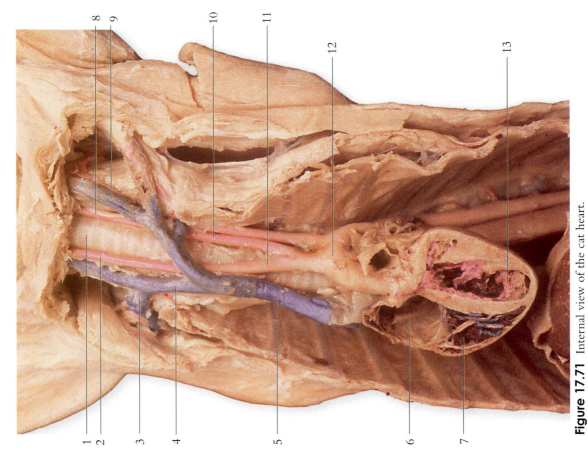

Figure 17.71 Internal view of the cat heart.

1. Trachea	8. Left common carotid a.
2. Right common carotid a.	9. Internal jugular v.
3. Right subclavian v.	10. Left common carotid a.
4. Right brachiocephalic v.	11. Right common carotid a.
5. Cranial (superior) vena cava	12. Aortic arch
6. Right atrium	13. Left ventricle
7. Right ventricle	

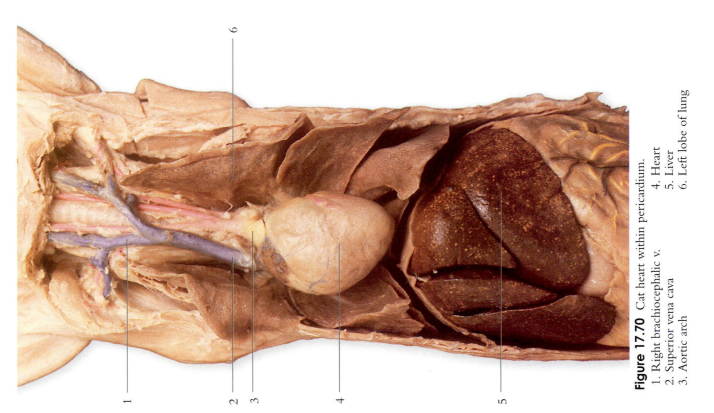

Figure 17.70 Cat heart within pericardium.

1. Right brachiocephalic v.	4. Heart
2. Superior vena cava	5. Liver
3. Aortic arch	6. Left lobe of lung

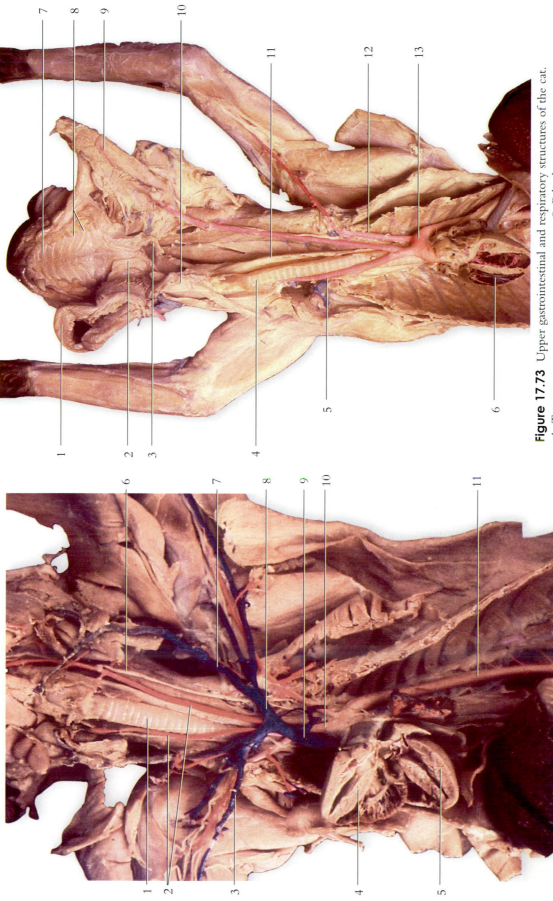

Figure 17.73 Upper gastrointestinal and respiratory structures of the cat.

1. Tongue
2. Soft palate
3. Epiglottis
4. Trachea
5. Right common carotid artery
6. Heart (cut)
7. Hard palate
8. Palatal rugae
9. Mandible (cut)
10. Larynx
11. Esophagus
12. Left subclavian artery
13. Aortic arch

Figure 17.72 Cat heart and surrounding structures.

1. Trachea
2. Common carotid arteries
3. Axillary vein
4. Heart (cut)
5. Left ventricle
6. Vagus nerve
7. External jugular vein
8. Left Brachiocephalic vein
9. Cranial (superior) vena cava
10. Brachiocephalic trunk
11. Thoracic aorta

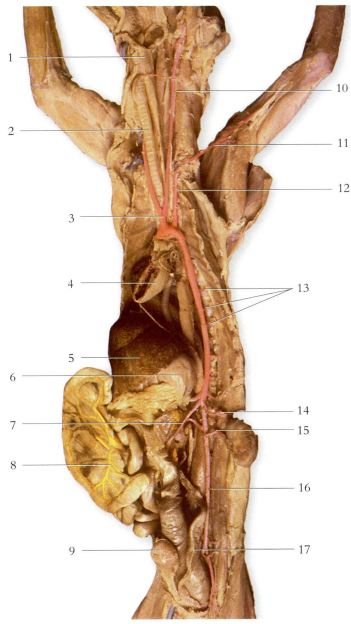

Figure 17.74 Anterior view of the arteries and veins of the trunk of a cat.

1. Larynx
2. Trachea
3. Brachiocephalic trunk
4. Heart (cut)
5. Liver
6. Stomach
7. Superior mesenteric a.
8. Superior mesenteric v.
9. Urinary bladder
10. Left common carotid a.
11. Left axillary a.
12. Left subclavian a.
13. Intercostal a.a.
14. Suprarenal a.
15. Renal a.
16. Abdominal aorta
17. Left horn of uterus

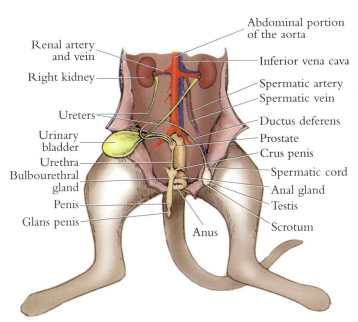

Figure 17.75 Diagram of the urogenital system of a male cat.

Renal artery and vein
Right kidney
Ureters
Urinary bladder
Urethra
Bulbourethral gland
Penis
Glans penis
Abdominal portion of the aorta
Inferior vena cava
Spermatic artery
Spermatic vein
Ductus deferens
Prostate
Crus penis
Spermatic cord
Anal gland
Testis
Scrotum
Anus

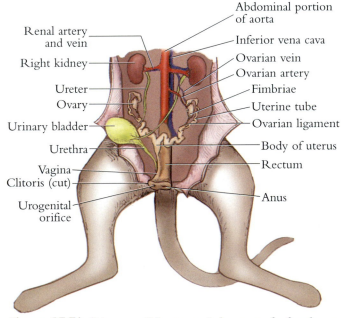

Figure 17.76 Diagram of the urogenital system of a female cat.

Renal artery and vein
Right kidney
Ureter
Ovary
Urinary bladder
Urethra
Vagina
Clitoris (cut)
Urogenital orifice
Abdominal portion of aorta
Inferior vena cava
Ovarian vein
Ovarian artery
Fimbriae
Uterine tube
Ovarian ligament
Body of uterus
Rectum
Anus

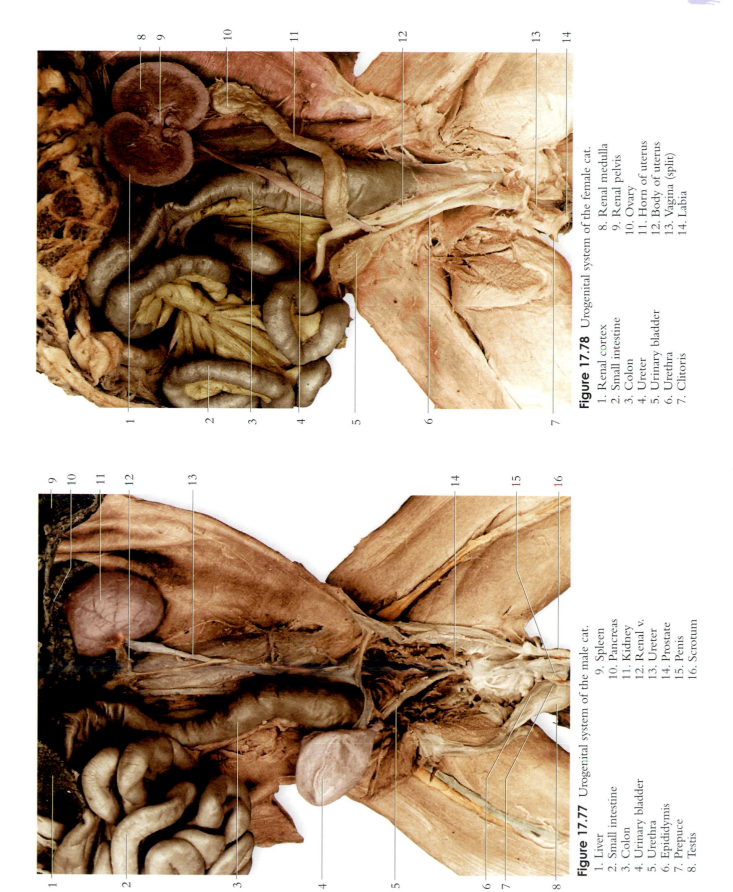

Figure 17.78 Urogenital system of the female cat.

1. Renal cortex
2. Small intestine
3. Colon
4. Ureter
5. Urinary bladder
6. Urethra
7. Clitoris
8. Renal medulla
9. Renal pelvis
10. Ovary
11. Horn of uterus
12. Body of uterus
13. Vagina (split)
14. Labia

Figure 17.77 Urogenital system of the male cat.

1. Liver
2. Small intestine
3. Colon
4. Urinary bladder
5. Urethra
6. Epididymis
7. Prepuce
8. Testis
9. Spleen
10. Pancreas
11. Kidney
12. Renal v.
13. Ureter
14. Prostate
15. Penis
16. Scrotum

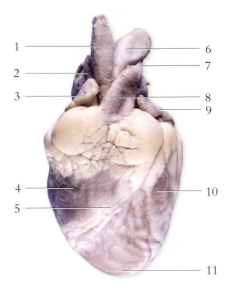

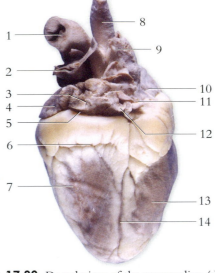

Figure 17.79 Ventral view of the mammalian (sheep) heart.

1. Brachiocephalic a.
2. Cranial vena cava
3. Right auricle of right atrium
4. Right ventricle
5. Interventicular groove
6. Aortic arch
7. Ligamentum arteriosum
8. Pulmonary trunk
9. Left auricle of left atrium
10. Left ventricle
11. Apex of heart

Figure 17.80 Dorsal view of the mammalian (sheep) heart.

1. Aorta
2. Pulmonary a.
3. Pulmonary v.
4. Left auricle
5. Left atrium
6. Atrioventricular groove
7. Left ventricle
8. Brachiocephalic a.
9. Cranial vena cava
10. Right auricle
11. Right atrium
12. Pulmonary v.
13. Right ventricle
14. Interventricular groove

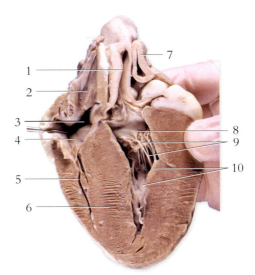

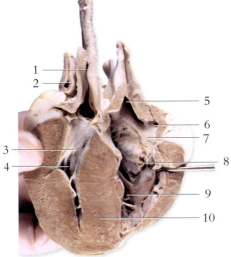

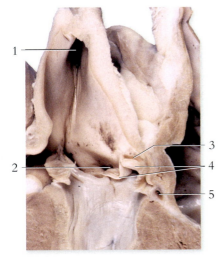

Figure 17.81 Coronal section of the mammalian (sheep) heart.

1. Aorta
2. Cranial vena cava
3. Right atrium
4. Right atrioventricular (tricuspid) valve
5. Right ventricle
6. Interventricalar septum
7. Pulmonary artery
8. Left atrioventricular (bicuspid) valve
9. Chordae tendineae
10. Papillary muscles

Figure 17.82 Coronal section of the mammalian (sheep) heart showing the valves.

1. Opening of the brachiocephalic a.
2. Pulmonary a.
3. Left atrioventricular (bicuspid) valve
4. Left ventricle
5. Opening of cranial vena cava
6. Opening of coronary sinus
7. Right atrium
8. Right atrioventricular (tricuspid) valve
9. Right ventricle
10. Interventricular septum

Figure 17.83 Coronal section of the mammalian (sheep) heart showing openings of coronary arteries.

1. Opening of brachiocephalic a.
2. Opening of left coronary a.
3. Opening of right coronary a.
4. Aortic valve
5. Coronary vessel

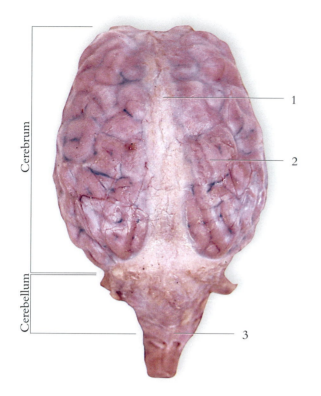

Figure 17.84 Sheep brain, dorsal view.
1. Dura mater covering longitudinal cerebral fissure
2. Arachnoid
3. Medulla oblongata

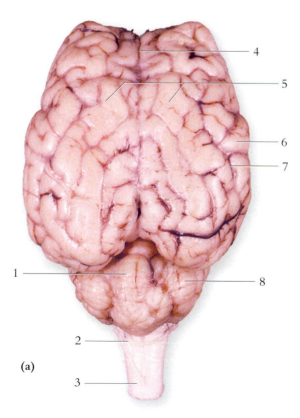

(a)

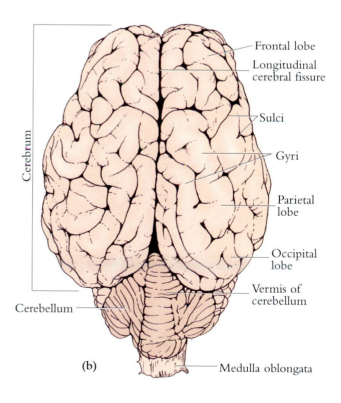

(b)

Frontal lobe
Longitudinal cerebral fissure
Sulci
Gyri
Parietal lobe
Occipital lobe
Vermis of cerebellum
Cerebellum
Medulla oblongata

Cerebrum

Figure 17.85 Sheep brain, dorsal view. (a) photograph; (b) diagram.
1. Vermis
2. Medulla oblongata
3. Spinal Cord
4. Longitudinal cerebral fissure
5. Cerebral hemipheres
6. Gyrus
7. Sulcus
8. Cerebellar hemisphere

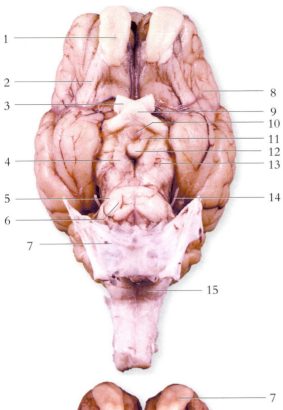

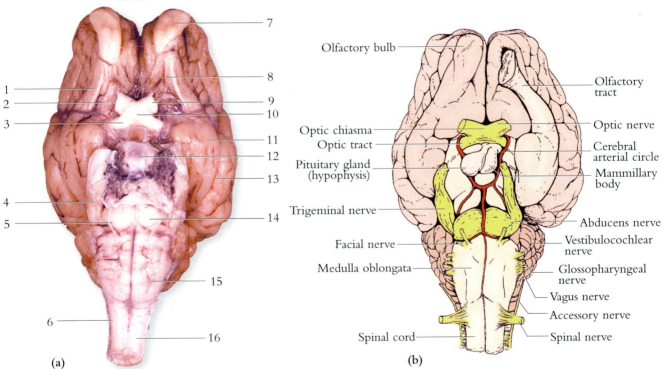

Figure 17.86 Ventral view of sheep brain with dura mater cut and reflected.
1. Olfactory bulb
2. Olfactory tract
3. Optic nerve
4. Oculomotor nerve
5. Trigeminal nerve
6. Pons
7. Dura mater (cut)
8. Pia mater (adhering to brain)
9. Optic chiasma
10. Position of pituitary stock
11. Tuber cinereum
12. Mammillary body
13. Cerebral penduncle
14. Trochlear nerve
15. Medulla oblongata

Figure 17.87 Sheep brain, venral view. (a) photograph; (b) diagram.
1. Lateral olfactory band
2. Olfactory trigone
3. Optic tract
4. Trigeminal nerve
5. Abducens nerve
6. Accessory nerve
7. Olfactory bulb
8. Medial olfactory band
9. Optic nerve
10. Optic chiasma
11. Pyriform lobe
12. Pituitary gland (hypophysis)
13. Rhinal sulcus
14. Pons
15. Medulla oblongata
16. Spinal cord

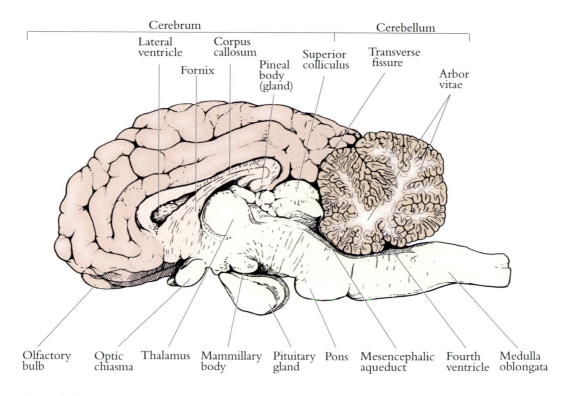

Figure 17.88 Sheep brain, sagittal view.

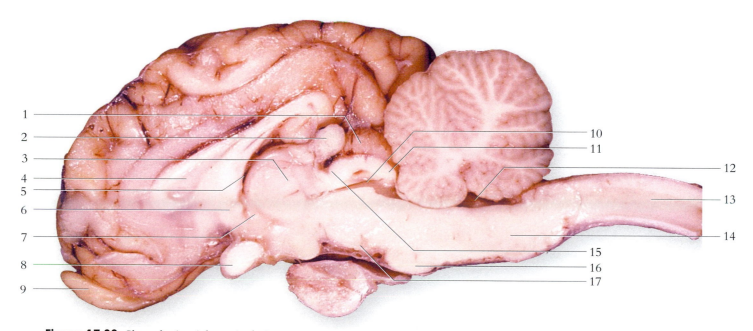

Figure 17.89 Sheep brain, right sagittal view.

1. Superior colliculus
2. Pineal body (gland)
3. Intermediate mass
4. Septum pellucidum
5. Interventicular foramen (foramen of Monro)
6. Anterior commissure
7. Third ventricle
8. Optic chiasma
9. Olfactory bulb
10. Mesencephalic (cerebral) aqueduct
11. Inferior colliculus
12. Fourth ventricle
13. Spinal cord
14. Medulla oblongata
15. Posterior commissure
16. Pons
17. Cerebral peduncle

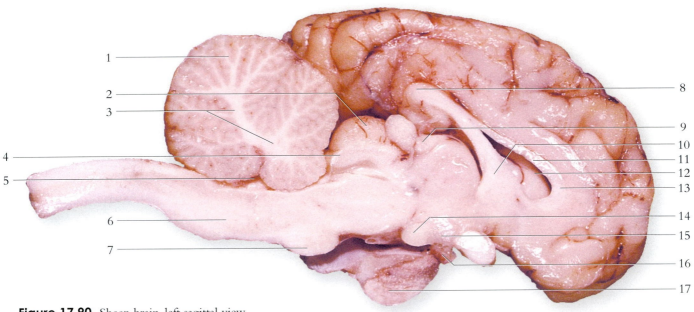

Figure 17.90 Sheep brain, left sagittal view.

1. Cerebellum
2. Superior colliculus
3. Arbor vitae
4. Inferior colliculus
5. Fourth ventricle
6. Medulla oblongata
7. Pons
8. Splenium of corpus callosum
9. Habenular trigone
10. Fornix
11. Body of corpus callosum
12. Lateral ventricle
13. Genu of corpus callosum
14. Mammillary body
15. Tuber cinereum
16. Pituitary stalk
17. Pituitary gland (hypophysis)

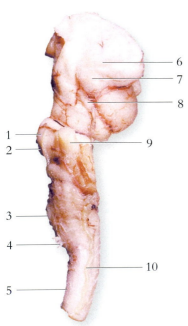

Figure 17.91 Cat brainstem, lateral view.
1. Pons
2. Abducens nerve
3. Medulla oblongata
4. Hypoglossal nerve
5. Spinal cord
6. Lateral geniculate body
7. Medial geniculate body
8. Trochlear nerve
9. Trigeminal nerve
10. Accessory nerve

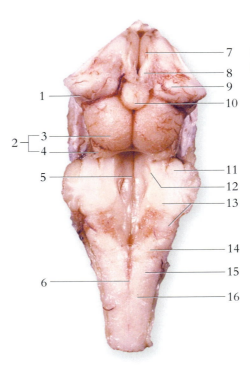

Figure 17.92 Cat brainstem, dorsal view.
1. Medial geniculate body
2. Corpora quadrigemina
3. Superior colliculus
4. Inferior colliculus
5. Fourth ventricle
6. Dorsal median sulcus
7. Intermediate mass
8. Habenular trigone
9. Thalamus
10. Pineal gland
11. Middle cerebellar peduncle
12. Anterior cerebellar penduncle
13. Posterior cerebellar peduncle
14. Tuberculum cuneatum
15. Fasciculus gracilis
16. Fasciculus cuneatus

Because humans are vertebrate organisms, the study of human biology is appropriate in a general zoology course. *Human anatomy* is the scientific discipline that investigates the structure of the body, and *human physiology* is the scientific discipline that investigates how body structures function. The purpose of this chapter is to present a visual overview of the principal anatomical structures of the human body.

Since both the *skeletal system* and the *muscular system* are concerned with body movement, they are frequently discussed together as the *skeletomusculature system*. In a functional sense, the flexible internal framework, or *bones* of the skeleton, support and provide movement at the *joints* where the muscles attached to the bones produce their actions as they are stimulated to contract.

The *nervous system* is anatomically divided into the *central nervous system* (CNS), which includes the *brain* and *spinal cord*, and the *peripheral nervous system* (PNS), which includes the *cranial nerves*, arising from the brain, and the *spinal nerves*, arising from the spinal cord. The *autonomic nervous system* (ANS) is a functional division of the nervous system devoted to regulation of involuntary activities of the body. The brain and spinal cord are the centers for integration and coordination of information. *Nerves*, composed of *neurons*, convey nerve impulses to and from the brain. *Sensory organs*, such as the eyes and ears, respond to impulses in the environment and convey sensations to the CNS. The nervous system functions with the *endocrine system* in coordinating body activities.

The *cardiovascular system* consists of the *heart, vessels* (both blood and lymphatic vessels), *blood*, and the tissues that produce the *blood*. The four-chambered human heart is enclosed with a *pericardial sac* within the thoracic cavity. *Arteries* and *arterioles* transport blood away from the heart, *capillaries* permeate the tissues and are the functional units for product exchange with the cells, and *venules* and *veins* transport blood toward the heart. *Lymphatic vessels* return interstitial fluid back to the circulatory system after first passing it through *lymph nodes* for cleansing. Blood cells are produced in the bone marrow and once old and worn, they are broken down in the liver.

The *respiratory system* consists of the *conducting division* that transports air to and from the *respiratory division* within the *lungs*. The *alveoli* of the lungs contact the capillaries of the cardiovascular system and are the sites for transport of respiratory gases into and out of the body.

The *digestive system* consists of a *gastrointestinal tract* (GI tract) and *accessory digestive organs*. Food traveling through the GI tract is processed such that it is suitable for absorption through the intestinal wall into the bloodstream. The *pancreas* and *liver* are the principal digestive organs. The pancreas produces hormones and enzymes. The liver processes nutrients, stores glucose as glycogen, and excretes bile.

Because of commonalty of prenatal development and dual functions of some of the organs, the *urinary system* and *reproductive system* may be considered together as the *urogenital system*. The urinary system, consisting of the *kidneys, ureters, urinary bladder*, and *urethra*, extracts and processes wastes from the blood in the form of urine. The male and female reproductive systems produce regulatory hormones and gametes (sperm and ova, respectively) within the gonads (testes and ovaries). Sexual reproduction is the mechanism for propagation of offspring that have traits from both parents. The process of prenatal development is made possible by the formation of *extraembryonic membranes* (placenta, umbilical cord, allantois, amnion, chorion, and yolk sac) within the uterus of the pregnant woman.

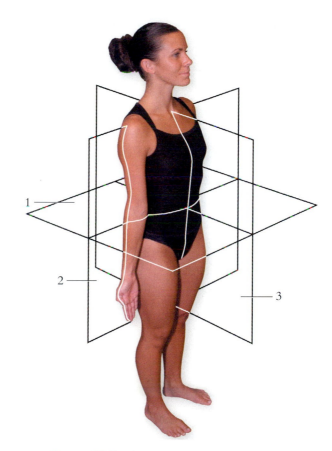

Figure 18.1 Planes of reference in a person while standing in anatomical position. The anatomical position provides a basis of reference for describing the relationship of one body part to another. In the anatomical position, the person is standing, the feet are parallel, the eyes are directed forward, and the arms are to the sides with the palms turned forward and the fingers are pointed straight down.

1. Transverse plane
   (cross-sectional plane)
2. Coronal plane
   (frontal plane)
3. Sagittal plane

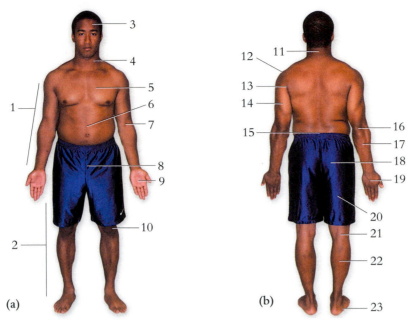

(a)   (b)

Figure 18.2 Major body parts and regions in humans (bipedal vertebrate).
(a) Anterior view and (b) a posterior view.

1. Upper extremity	9. Palmar region (palm)	17. Antebrachium (forearm)
2. Lower extremity	10. Patellar region (patella)	18. Gluteal region (buttock)
3. Head	11. Cervical region	19. Dorsum of hand
4. Neck, anterior aspect	12. Shoulder	20. Thigh
5. Thorax (chest)	13. Axilla (armpit)	21. Popliteal fossa
6. Abdomen	14. Brachium (upper arm)	22. Calf
7. Cubital fossa	15. Lumbar region	23. Plantar surface (sole)
8. Pubic region	16. Elbow	

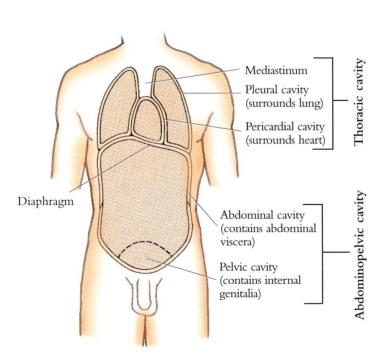

Mediastinum

Pleural cavity
(surrounds lung)

Pericardial cavity
(surrounds heart)

Thoracic cavity

Diaphragm

Abdominal cavity
(contains abdominal
viscera)

Pelvic cavity
(contains internal
genitalia)

Abdominopelvic cavity

Figure 18.3 Anterior view of (coronal plane)
the body cavities of the trunk.

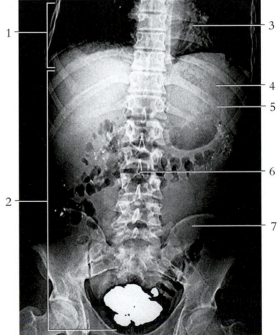

Figure 18.4 MR image of the trunk showing
the body cavities and their contents.

1. Thoracic cavity	5. Image of rib
2. Abdominopelvic cavity	6. Image of lumbar
3. Image of heart	vertebra
4. Image of diaphragm	7. Image of Ilium

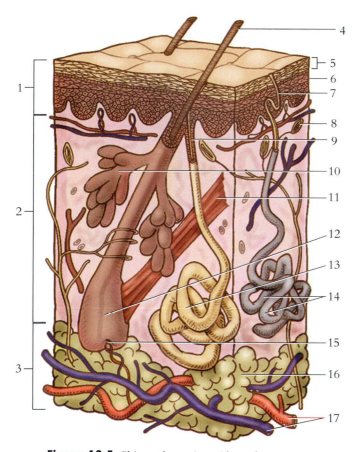

Figure 18.5 Skin and certain epidermal structures.

1. Epidermis
2. Dermis
3. Hypodermis
4. Shaft of hair
5. Stratum corneum
6. Stratum basale
7. Sweat duct
8. Sensory receptor
9. Sweat duct
10. Sebaceous gland
11. Arrector pili muscle
12. Hair follicle
13. Apocrine sweat gland
14. Eccrine sweat gland
15. Bulb of hair
16. Adipose tissue
17. Cutaneous blood vessels

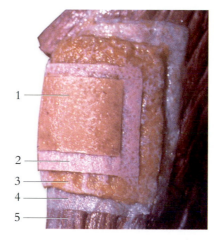

Figure 18.6 Gross structure of the skin and underlying fascia.

1. Epidermis
2. Dermis
3. Hypodermis
4. Fascia
5. Muscle

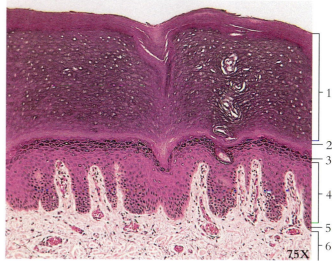

Figure 18.7 Epidermis and dermis.

1. Stratum corneum
2. Stratum lucidum
3. Stratum granulosum
4. Stratum spinosum
5. Stratum basale
6. Dermis

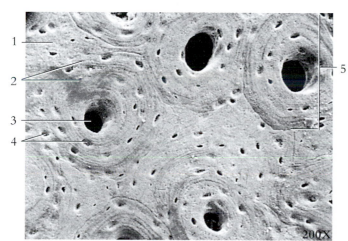

Figure 18.8 Electron micrograph of bone tissue.

1. Interstitial lamellae
2. Lamellae
3. Central canal (haversian canal)
4. Lacunae
5. Osteon (haversian system)

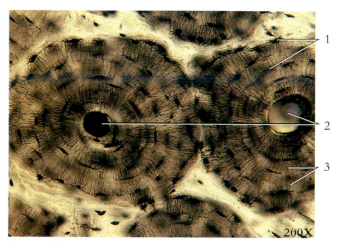

Figure 18.9 Transverse section of two osteons.

1. Lacunae
2. Central (haversian) canals
3. Lamellae

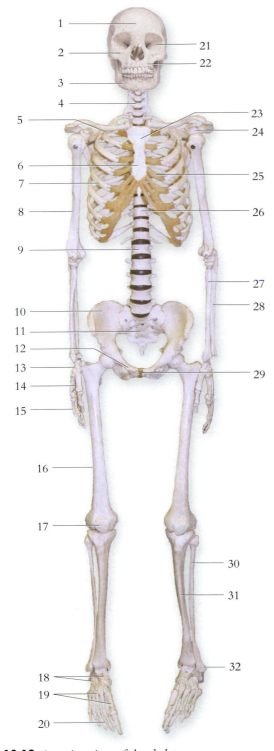

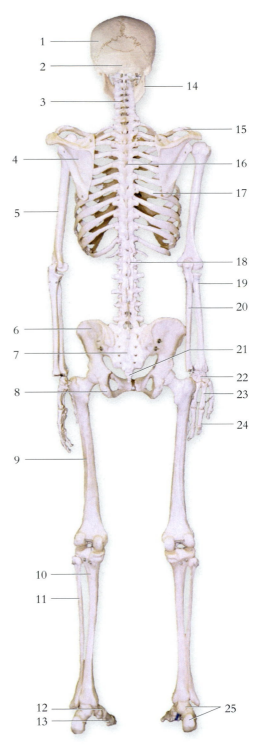

Figure 18.10 Anterior view of the skeleton.

1. Frontal bone
2. Zygomatic bone
3. Mandible
4. Cervical vertebra
5. Clavicle
6. Body of sternum
7. Rib
8. Humerus
9. Lumbar vertebra
10. Ilium
11. Sacrum
12. Pubis
13. Carpal bones
14. Metacarpal bones
15. Phalanges
16. Femur
17. Patella
18. Tarsal bones
19. Metatarsal bones
20. Phalanges
21. Orbit
22. Maxilla
23. Scapula
24. Manubrium
25. Costal cartilage
26. Thoracic vertebra
27. Radius
28. Ulna
29. Symphysis pubis
30. Fibula
31. Tibia
32. Calcaneus

Figure 18.11 Posterior view of the skeleton.

1. Parietal bone
2. Occipital bone
3. Cervical vertebra
4. Scapula
5. Humerus
6. Ilium
7. Sacrum
8. Ischium
9. Femur
10. Tibia
11. Fibula
12. Metatarsal bones
13. Phalanges
14. Mandible
15. Clavicle
16. Thoracic vertebra
17. Rib
18. Lumbar vertebra
19. Radius
20. Ulna
21. Coccyx
22. Carpal bones
23. Metacarpal bones
24. Phalanges
25. Tarsal bones

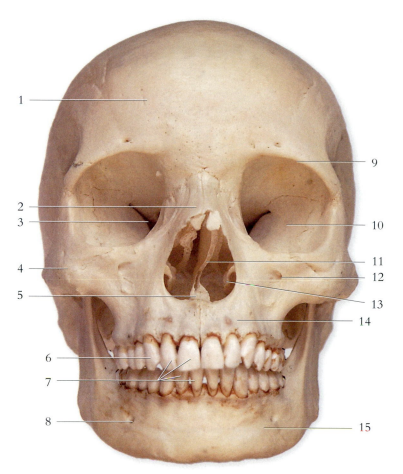

Figure 18.12 Anterior view of the human skull.
1. Frontal bone
2. Nasal bone
3. Superior orbital fissure
4. Zygomatic bone
5. Vomer
6. Canine
7. Incisors
8. Mental foramen
9. Supraorbital margin
10. Sphenoid bone
11. Perpendicular plate of ethmoid bone
12. Infraorbital foramen
13. Inferior nasal concha
14. Maxilla
15. Mandible

Figure 18.13 Lateral view of the human skull.
1. Coronal suture
2. Frontal bone
3. Lacrimal bone
4. Nasal bone
5. Zygomatic bone
6. Maxilla
7. Premolars
8. Molars
9. Mandible
10. Parietal bone
11. Squamosal suture
12. Temporal bone
13. Lambdoidal suture
14. External acoustic meatus
15. Occipital bone
16. Condylar process of mandible
17. Mandibular notch
18. Mastoid process of temporal bone
19. Coronoid process of mandible
20. Angle of mandible

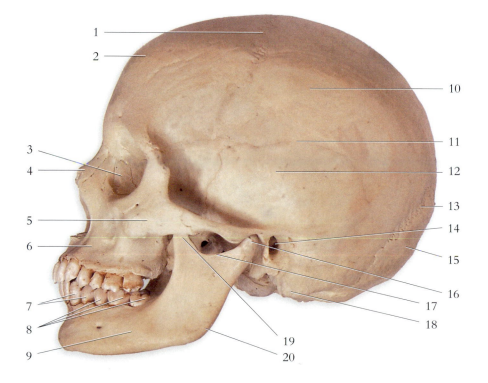

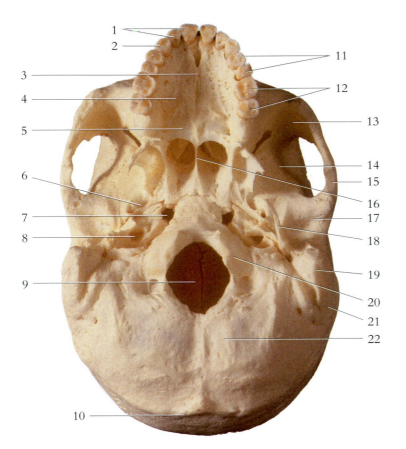

Figure 18.14 Inferior view of the human skull.

1. Incisors
2. Canine
3. Intermaxillary suture
4. Maxilla
5. Palatine bone
6. Foramen ovale
7. Foramen lacerum
8. Carotid canal
9. Foramen magnum
10. Superior nuchal line
11. Premolars
12. Molars
13. Zygomatic bone
14. Sphenoid bone
15. Zygomatic arch
16. Vomer
17. Mandibular fossa
18. Styloid process of temporal bone
19. Mastoid process of temporal bone
20. Occipital condyle
21. Temporal bone
22. Occipital bone

Figure 18.15 Sagittal view of the human skull.

1. Frontal bone
2. Frontal sinus
3. Crista galli of ethmoid bone
4. Cribriform plate of ethmoid bone
5. Nasal bone
6. Nasal concha
7. Maxilla
8. Mandible
9. Parietal bone
10. Occipital bone
11. Internal acoustic meatus
12. Sella turcica
13. Hypoglossal canal
14. Sphenoidal sinus
15. Styloid process of temporal bone
16. Vomer

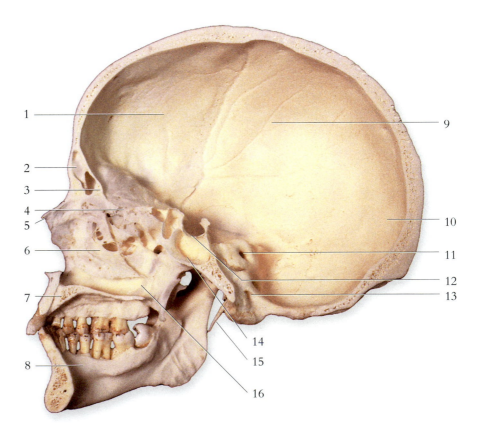

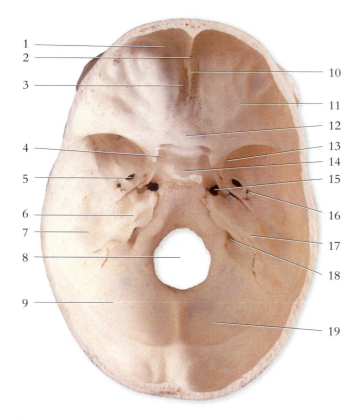

Figure 18.16 Superior view of the cranium.

1. Frontal bone
2. Foramen cecum
3. Cribriform plate of ethmoid bone
4. Optic canal
5. Foramen ovale
6. Petrous part of temporal bone
7. Temporal bone
8. Foramen magnum
9. Occipital bone
10. Crista galli of ethmoid bone
11. Anterior cranial fossa
12. Sphenoid bone
13. Foramen rotundum
14. Sella turcica of sphenoid bone
15. Foramen lacerum
16. Foramen spinosum
17. Internal acoustic meatus
18. Jugular foramen
19. Posterior cranial fossa

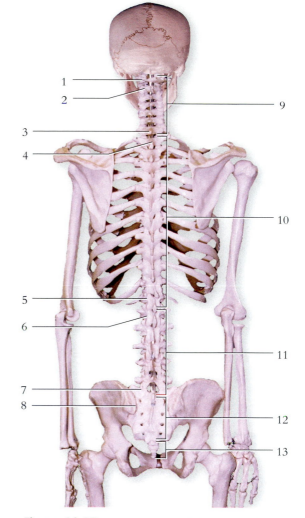

Figure 18.17 Posterior view of the vertebral column.

1. Atlas
2. Axis
3. Seventh cervical vertebra
4. First thoracic vertebra
5. Twelfth thoracic vertebra
6. First lumbar vertebra
7. Fifth lumbar vertebra
8. Sacroiliac joint
9. Cervical vertebrae
10. Thoracic vertebrae
11. Lumbar vertebrae
12. Sacrum
13. Coccyx

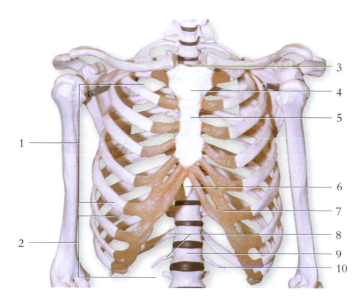

Figure 18.18 Anterior view of the rib cage.

1. True ribs (seven pairs)
2. False ribs (five pairs)
3. Jugular notch
4. Manubrium
5. Body of sternum
6. Xiphoid process
7. Costal cartilage
8. Floating ribs (inferior two pairs of false ribs)
9. Twelfth thoracic vertebra
10. Twelfth rib

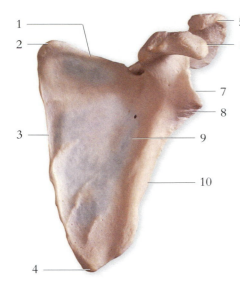

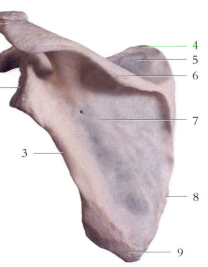

Figure 18.19 Anterior view of the left scapula.
1. Superior border
2. Superior angle
3. Medial (vertebral) border
4. Inferior angle
5. Acromion
6. Coracoid process
7. Glenoid fossa
8. Infraglenoid tubercle
9. Subscapular fossa
10. Lateral (axillary) border

Figure 18.20 Posterior view of the left scapula.
1. Acromion
2. Glenoid fossa
3. Lateral (axillary) border
4. Superior angle
5. Supraspinous fossa
6. Spine
7. Infraspinous fossa
8. Medial (vertebral) border
9. Inferior angle

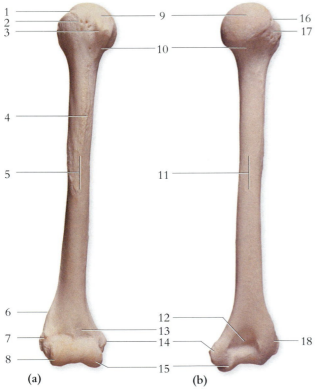

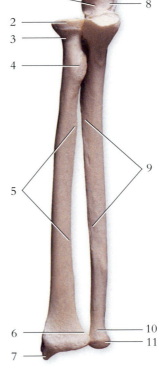

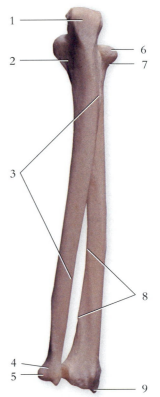

Figure 18.21 Right humerus. (a) Anterior view (b) Posterior view.
1. Greater tubercle
2. Intertubercular groove
3. Lesser tubercle
4. Deltoid tuberosity
5. Anterior body (shaft) of humerus
6. Lateral supracondylar ridge
7. Lateral epicondyle
8. Capitulum
9. Head of humerus
10. Surgical neck
11. Posterior body (shaft) of humerus
12. Olecranon fossa
13. Coronoid fossa
14. Medial epicondyle
15. Trochlea
16. Anatomical neck
17. Greater tubercle
18. Lateral epicondyle

Figure 18.22 Anterior view of the right ulna and radius.
1. Trochlear notch
2. Head of radius
3. Neck of radius
4. Radial tuberosity
5. Interosseous margin
6. Ulnar notch of radius
7. Styloid process of radius
8. Olecranon
9. Interosseous margin
10. Neck of ulna
11. Head of ulna

Figure 18.23 Posterior view of the right ulna and radius.
1. Olecranon
2. Radial notch of ulna
3. Interosseous margin
4. Head of ulna
5. Styloid process of ulna
6. Head of radius
7. Neck of radius
8. Interosseous margin
9. Styloid process of radius

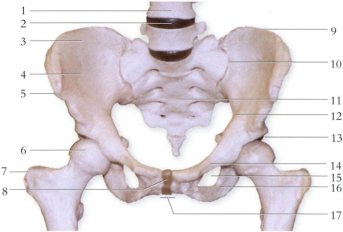

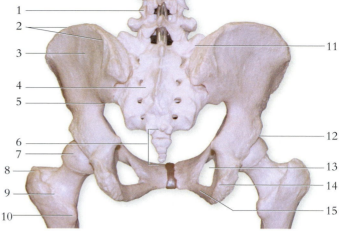

Figure 18.24 Anterior view of the articulated pelvic girdle showing the two coxal bones, the sacrum, and the two femora.

1. Lumbar vertebra
2. Intervertebral disc
3. Ilium
4. Iliac fossa
5. Anterior superior iliac spine
6. Head of femur
7. Greater trochanter
8. Symphysis pubis
9. Crest of the ilium
10. Sacroiliac joint
11. Sacrum
12. Pelvic brim
13. Acetabulum
14. Pubic crest
15. Obturator foramen
16. Ischium
17. Pubic angle

Figure 18.25 Posterior view of the articulated pelvic girdle showing the two coxal bones, the sacrum, and the two femora.

1. Lumbar vertebra
2. Crest of ilium
3. Ilium
4. Sacrum
5. Greater sciatic notch
6. Coccyx
7. Head of femur
8. Greater trochanter
9. Intertrochanteric crest
10. Lesser trochanter
11. Sacroiliac joint
12. Acetabulum
13. Obturator foramen
14. Ischium
15. Pubis

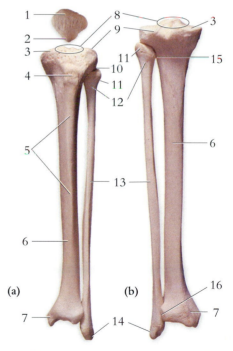

Figure 18.26 Left femur. (a) An anterior view and (b) a posterior view.

1. Fovea capitis femoris
2. Head
3. Neck
4. Lesser trochanter
5. Medial epicondyle
6. Patellar surface
7. Greater trochanter
8. Intertrochanteric crest
9. Intertrochanteric line
10. Lateral epicondyle
11. Lateral condyle
12. Intercondylar fossa
13. Head
14. Fovea capitis femoris
15. Neck
16. Lesser trochanter
17. Linea aspera on shaft (body) of femur
18. Medial epicondyle
19. Medial condyle

Figure 18.27 Anterior view of the (a) left patella, tibia, and fibula. (b) A posterior view of the left tibia and fibula.

1. Base of patella
2. Apex of patella
3. Medial condyle
4. Tibial tuberosity
5. Anterior crest of tibia
6. Body (shaft) of tibia
7. Medial malleolus
8. Intercondylar tubercles
9. Lateral condyle
10. Tibial articular facet of fibula
11. Head of fibula
12. Neck of fibula
13. Body (shaft) of fibula
14. Lateral malleolus
15. Fibular articular facet of tibia
16. Fibular notch of tibia

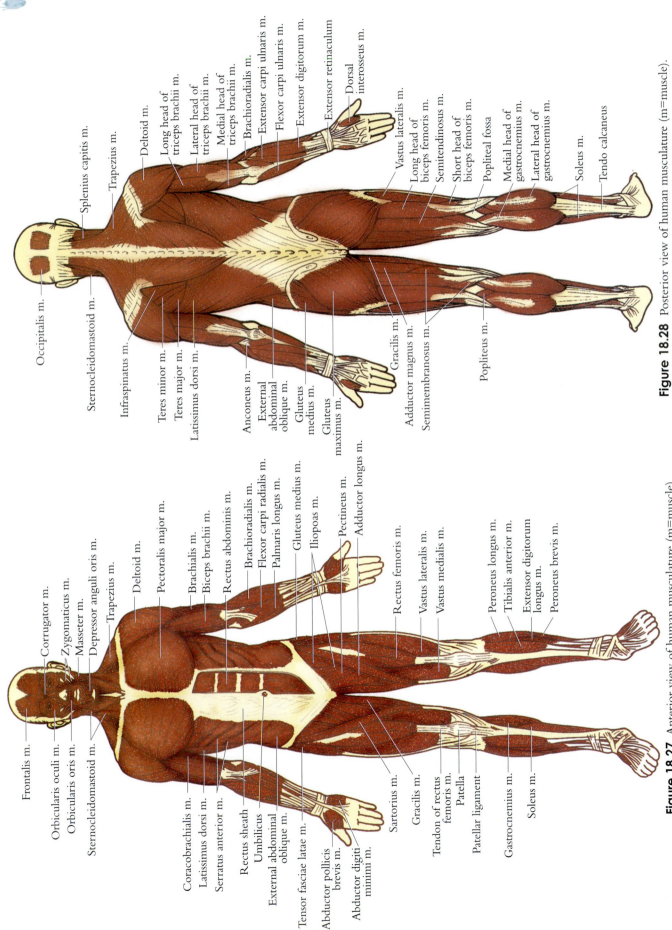

Figure 18.28 Posterior view of human musculature (m=muscle).

Figure 18.27 Anterior view of human musculature (m=muscle).

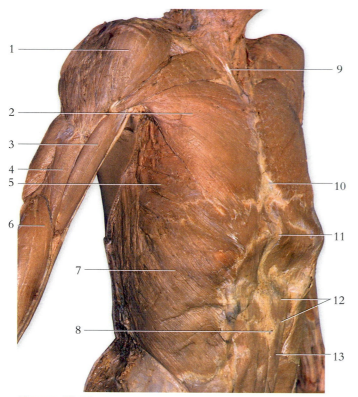

Figure 18.29 Anterolateral view of the trunk.

1. Deltoid m.
2. Pectoralis major m.
3. Biceps brachii m. (long head)
4. Brachialis m.
5. Serratus anterior m.
6. Brachioradialis m.
7. External abdominal oblique m.
8. Umbilicus
9. Tendon of sternocleidomastoid m.
10. Sternum.
11. Xiphoid process
12. Tendinous inscriptions of rectus abdominis m.
13. Rectus abdominis m.

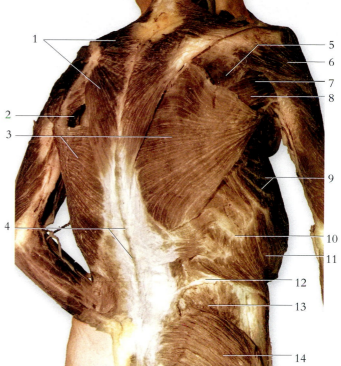

Figure 18.30 Posterolateral view of the trunk.

1. Trapezius m.
2. Triangle of ausculation
3. Latissimus dorsi mm.
4. Vertebral column (spinous processes)
5. Infraspinatus m.
6. Deltoid m.
7. Teres minor m.
8. Teres major m.
9. Serratus anterior mm.
10. Rib
11. External abdominal oblique m.
12. Iliac crest
13. Gluteus medius m.
14. Gluteus maximus m.

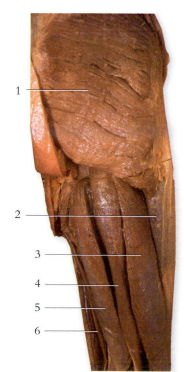

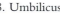

Figure 18.31 Superficial muscles of gluteal and thigh regions.

1. Gluteus maximus m.
2. Vastus lateralis m.
3. Biceps femoris m.
4. Semitendinosus m.
5. Semimembranosus m.
6. Gracilis m.

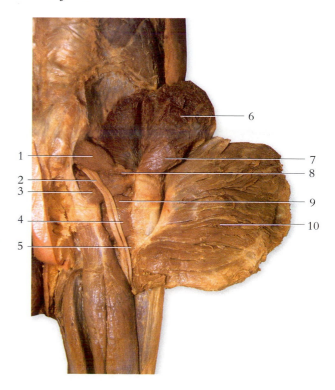

Figure 18.32 Deep structures of gluteal region.

1. Piriformis m.
2. Sciatic n.
3. Obturator internus m.
4. Quadratus femoris m.
5. Adductor minimus m.
6. Gluteus medius m.
7. Gluteus minimus m.
8. Superior gemellus m.
9. Inferior gemellus m.
10. Gluteus maximus m.

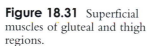

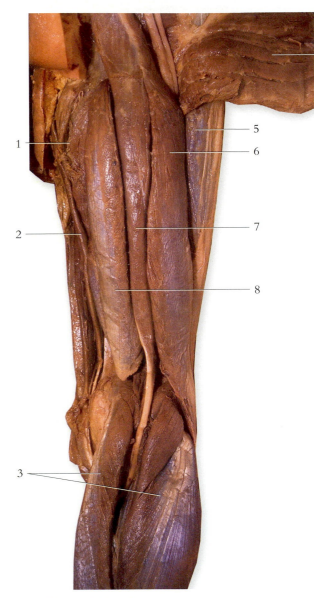

Figure 18.33 Deep posterior structures
of thigh and popliteal regions.
1. Adductor magnus m.
2. Gracilis m.
3. Gastrocnemius m.
4. Gluteus maximus m.
5. Vastus lateralis m.
6. Biceps femoris m.
7. Semitendinosus m.
8. Semimembranosus m.

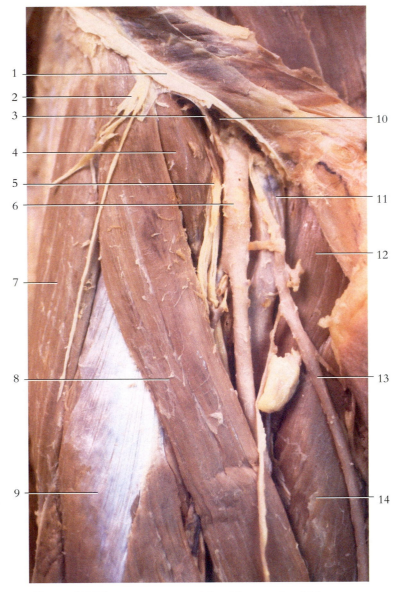

Figure 18.34 Anterior view of the right superior thigh.
1. Inguinal ligament
2. Lateral femoral
   cutaneous nerve
3. Superficial circumflex
   iliac artery
4. Iliopsoas m.
5. Femoral nerve
6. Femoral artery
7. Tensor fasciae latae m.
8. Sartorius m.
9. Rectus femoris m.
10. Femoral ring
11. Femoral vein
12. Pectineus m.
13. Great saphenous vein
14. Adductor longus m.

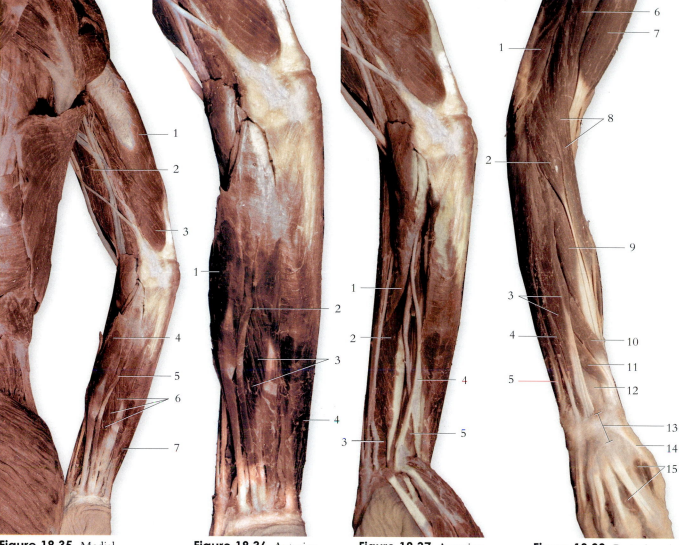

Figure 18.35 Medial brachium and superficial flexors of the right antebrachium.
1. Triceps brachii m. (lateral head)
2. Biceps brachii m. (short head)
3. Triceps brachii m. (medial head)
4. Flexor carpi radialis m.
5. Palmaris longus m.
6. Superficial digital flexor m.
7. Flexor carpi ulnaris m.

Figure 18.36 Anterior view of the superficial muscles of the right forearm.
1. Flexor carpi radialis m.
2. Palmaris longus m.
3. Superficial digital flexor m.
4. Flexor carpi ulnaris m.

Figure 18.37 Anterior view of the deep muscles of the right forearm.
1. Pronator teres m.
2. Flexor pollicis longus m.
3. Pronator quadratus m.
4. Median nerve
5. Deep digital flexor m.

Figure 18.38 Posterior view of the superficial muscles of the right forearm.
1. Triceps brachii m. (medial head)
2. Extensor carpi radialis longus m.
3. Extensor digitorum m.
4. Extensor digiti minimi m.
5. Extensor carpi ulnaris m.
6. Brachialis m.
7. Biceps brachii m. (long head)
8. Brachioradialis m.
9. Extensor carpi radialis brevis m.
10. Abductor pollicis longus m.
11. Extensor pollicis brevis m.
12. Radius
13. Extensor retinaculum
14. Tendon of extensor pollicis longus m.
15. Dorsal interosseous mm.

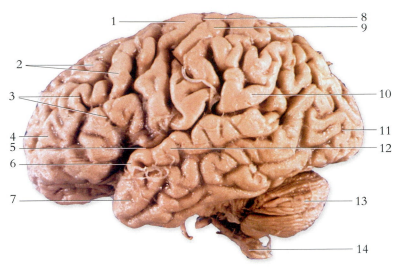

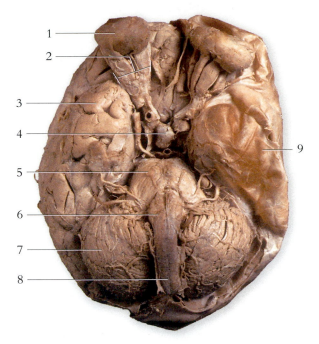

Figure 18.39 Lateral view of the brain.

1. Primary motor
   cerebral cortex
2. Gyri
3. Sulci
4. Frontal lobe of cerebrum
5. Lateral sulcus
6. Olfactory cerebral cortex
7. Temporal lobe of cerebrum
8. Central sulcus
9. Primary sensory
   cerebral cortex
10. Parietal lobe of cerebrum
11. Occipital lobe of cerebrum
12. Auditory cerebral cortex
13. Cerebellum
14. Medulla oblongata

Figure 18.40 Inferior view of the brain with the eyes and part of the meninges still intact.

1. Eyeball
2. Muscles of the eye
3. Temporal lobe of cerebrum
4. Pituitary gland
5. Pons
6. Medulla oblongata
7. Cerebellum
8. Spinal cord
9. Dura mater

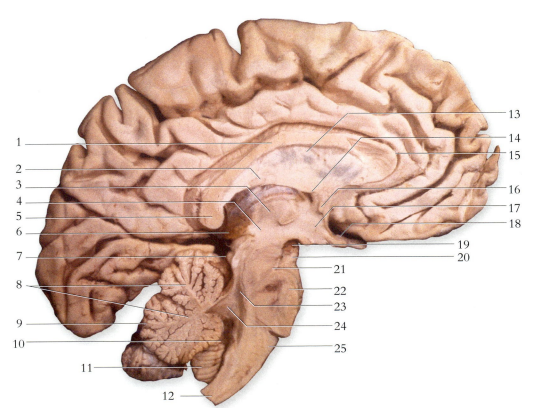

Figure 18.41 Sagittal view of the brain.

1. Body of corpus callosum
2. Crus of fornix
3. Third ventricle
4. Posterior commissure
5. Splenium of corpus callosum
6. Pineal body
7. Inferior colliculus
8. Arbor vitae of cerebellum
9. Vermis of cerebellum
10. Choroid plexus of fourth
    ventricle
11. Tonsilla of cerebellum
12. Medulla oblongata
13. Septum pellucidum (cut)
14. Intraventricular foramen
15. Genu of corpus callosum
16. Anterior commissure
17. Hypothalmus
18. Optic chiasma
19. Oculomotor nerve
20. Cerebral peduncle
21. Midbrain
22. Pons
23. Mesencephalic
    (cerebral) aqueduct
24. Fourth ventricle
25. Pyramid of medulla oblongata

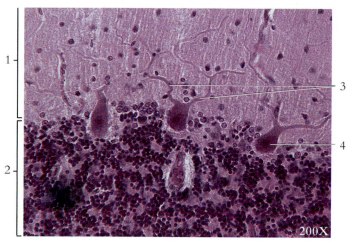

Figure 18.42 Purkinje neurons from the cerebellum.
1. Molecular layer of cerebellar cortex
2. Granular layer of cerebellar cortex
3. Dendrites of Purkinje cell
4. Purkinje cell body

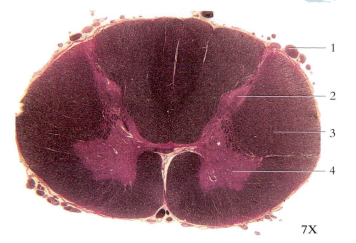

Figure 18.43 Transverse section of the spinal cord.
1. Posterior (dorsal) root of spinal nerve
2. Posterior (dorsal) horn (gray matter)
3. Spinal cord tract (white matter)
4. Anterior (ventral) horn (gray matter)

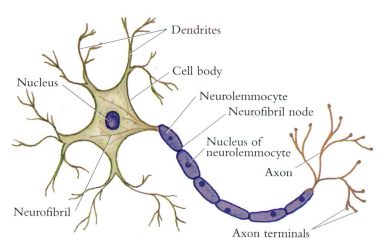

Figure 18.44 Structure of a myelinated neuron.

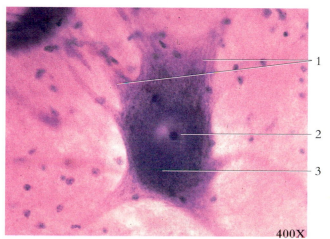

Figure 18.45 Photomicrograph of a neuron.
1. Cytoplasmic extensions   3. Cell body of neuron
2. Nucleus

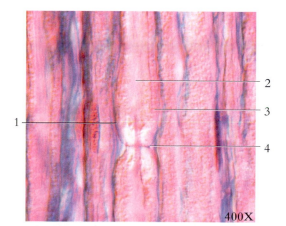

Figure 18.46 Histology of a myelinated nerve.
1. Neurolemmal sheath   4. Neurofibril node
2. Axon   (node of Ranvier)
3. Myelin layer

Figure 18.47 Transverse section of a nerve.
1. Perineurium   3. Endoneurium
2. Epineurium   4. Bundle of axons

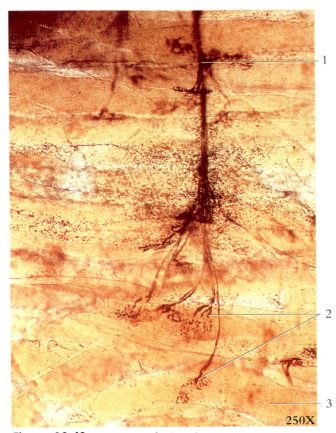

Figure 18.48 Neuromuscular junction.
1. Motor nerve   3. Skeletal muscle fiber
2. Motor end plates

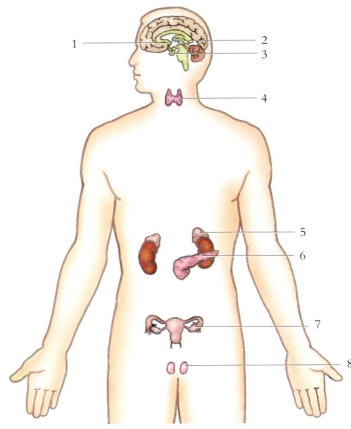

Figure 18.49 Principal endocrine glands.
1. Hypothalamus   5. Adrenal (suprarenal) gland
2. Pineal body   6. Pancreas
3. Pituitary gland   7. Ovary
4. Thyroid and parathyroid glands   8. Testis

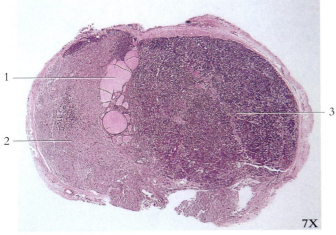

Figure 18.50 Pituitary gland.
1. Pars intermedia (adenohypophysis)
2. Pars nervosa (neurohypophysis)
3. Pars distalis (adenohypophysis)

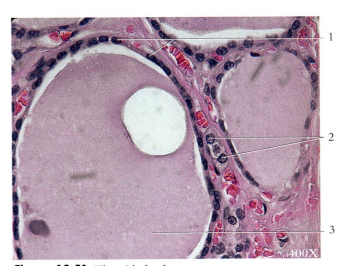

Figure 18.51 Thyroid gland.
1. Follicle cells
2. C cells
3. Colloid within follicle

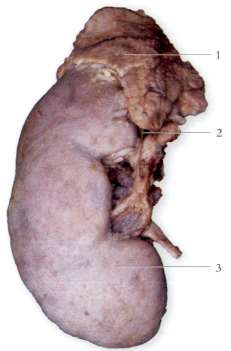

Figure 18.52 Adrenal (suprarenal) gland.
1. Adrenal gland
2. Inferior suprarenal artery
3. Kidney

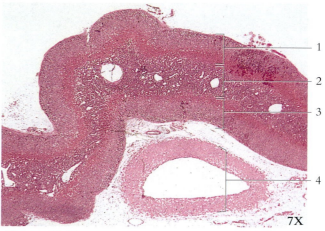

Figure 18.53 Adrenal gland.
1. Adrenal cortex
2. Adrenal medulla
3. Adrenal cortex
4. Blood vessel

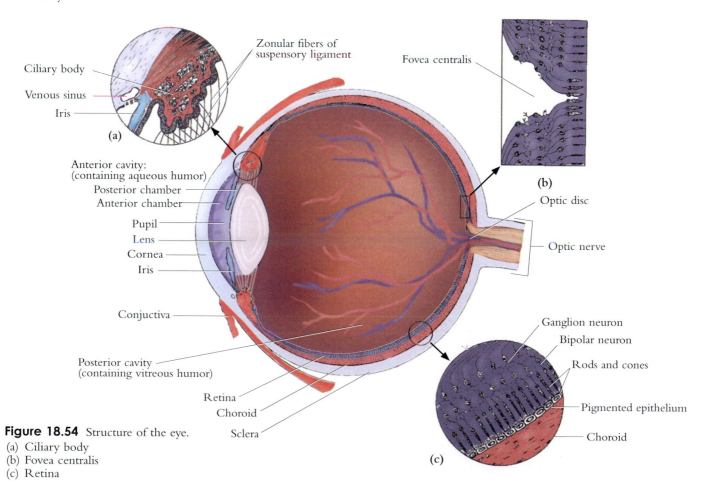

Zonular fibers of
suspensory ligament

Ciliary body

Venous sinus

Iris

(a)

Fovea centralis

(b)

Optic disc

Optic nerve

Anterior cavity:
(containing aqueous humor)

Posterior chamber

Anterior chamber

Pupil

Lens

Cornea

Iris

Conjuctiva

Posterior cavity
(containing vitreous humor)

Retina

Choroid

Sclera

Ganglion neuron

Bipolar neuron

Rods and cones

Pigmented epithelium

Choroid

(c)

Figure 18.54 Structure of the eye.
(a) Ciliary body
(b) Fovea centralis
(c) Retina

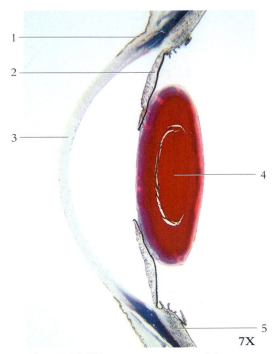

7X

Figure 18.55 Anterior portion of the eye.
1. Conjunctivia   4. Lens
2. Iris   5. Ciliary body
3. Cornea

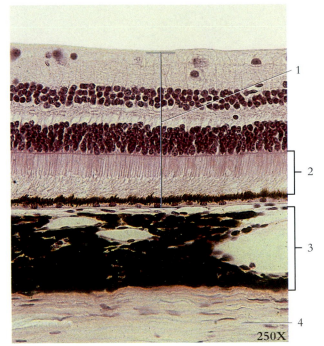

250X

Figure 18.56 Retina.
1. Retina   3. Choroid
2. Rods and cones 4. Sclera

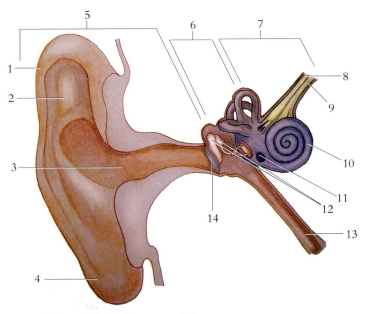

Figure 18.57 Structure of the ear.
1. Helix   8. Facial nerve
2. Auricle   9. Vestibulocochlear nerve
3. External auditory canal 10. Cochlea
4. Earlobe   11. Vestibular (oval) window
5. Outer ear   12. Auditory ossicles
6. Middle ear   13. Auditory tube
7. Inner ear   14. Tympanic membrane

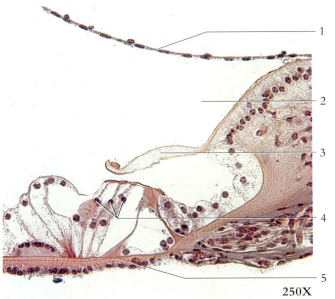

250X

Figure 18.58 Spiral organ (organ of Corti).
1. Vestibular membrane
2. Cochlear duct
3. Tectorial membrane
4. Hair cells
5. Basilar membrane

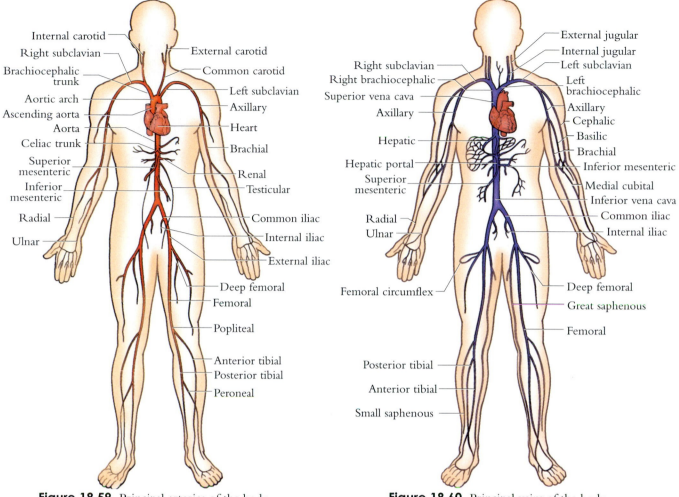

Figure 18.59 Principal arteries of the body.

Internal carotid
Right subclavian
Brachiocephalic trunk
Aortic arch
Ascending aorta
Aorta
Celiac trunk
Superior mesenteric
Inferior mesenteric
Radial
Ulnar

External carotid
Common carotid
Left subclavian
Axillary
Heart
Brachial
Renal
Testicular
Common iliac
Internal iliac
External iliac
Deep femoral
Femoral
Popliteal
Anterior tibial
Posterior tibial
Peroneal

Figure 18.60 Principal veins of the body.

Right subclavian
Right brachiocephalic
Superior vena cava
Axillary
Hepatic
Hepatic portal
Superior mesenteric
Radial
Ulnar
Femoral circumflex

External jugular
Internal jugular
Left subclavian
Left brachiocephalic
Axillary
Cephalic
Basilic
Brachial
Inferior mesenteric
Medial cubital
Inferior vena cava
Common iliac
Internal iliac
Deep femoral
Great saphenous
Femoral
Posterior tibial
Anterior tibial
Small saphenous

Figure 18.61 Position of the heart within the pericardium.

1. Mediastinum
2. Right lung
3. Pericardium
4. Diaphragm
5. Liver
6. Left lung

Figure 18.62 Anterior view of the heart and associated structures

1. Right vagus nerve
2. Right brachiocephalic vein
3. Superior vena cava
4. Right phrenic nerve
5. Ascending aorta
6. Pericardium (cut)
7. Right ventricle of heart
8. Brachiocephalic artery
9. Left brachiocephalic vein
10. Aortic arch
11. Left phrenic nerve
12. Left ventricle of heart
13. Apex of heart

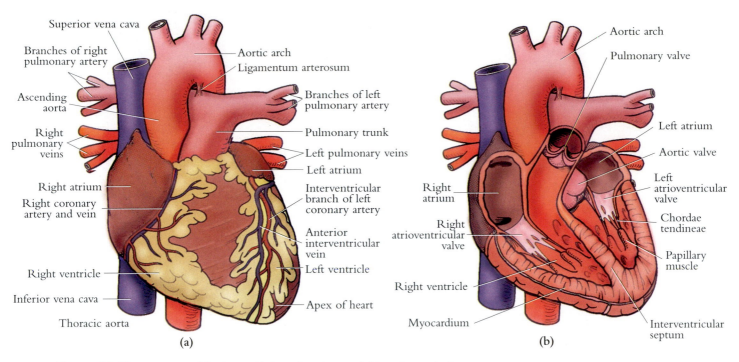

Superior vena cava

Branches of right
pulmonary artery

Ascending
aorta

Right
pulmonary
veins

Right atrium

Right coronary
artery and vein

Right ventricle

Inferior vena cava

Thoracic aorta

Aortic arch

Ligamentum arterosum

Branches of left
pulmonary artery

Pulmonary trunk

Left pulmonary veins

Left atrium

Interventricular
branch of left
coronary artery

Anterior
interventricular
vein

Left ventricle

Apex of heart

(a)

Aortic arch

Pulmonary valve

Left atrium

Aortic valve

Left
atrioventricular
valve

Chordae
tendineae

Papillary
muscle

Right
atrium

Right
atrioventricular
valve

Right ventricle

Myocardium

Interventricular
septum

(b)

Figure 18.63 Structure of the heart. (a) Anterior view and (b) an internal view.

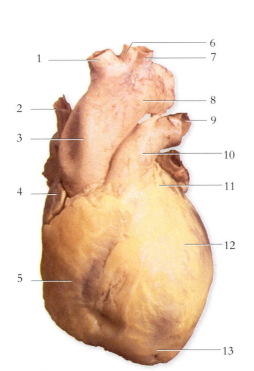

Figure 18.64 Anterior view of the heart and great vessels.
1. Brachiocephalic trunk
2. Superior vena cava
3. Ascending aorta
4. Right atrium
5. Right ventricle
6. Left common carotid artery
7. Left subclavian artery
8. Aortic arch
9. Pulmonary artery
10. Pulmonary trunk
11. Left atrium
12. Left ventricle
13. Apex of heart

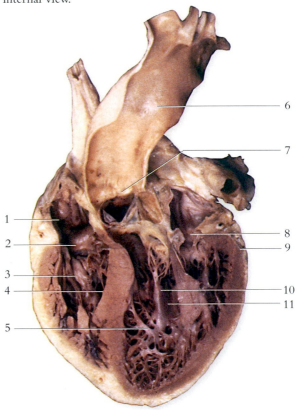

Figure 18.65 Internal structure of the heart.
1. Right atrium
2. Right atrioventricular valve
3. Right ventricle
4. Interventricular septum
5. Trabeculae carneae
6. Ascending aorta
7. Aortic valve
8. Left atrioventricular valve
9. Myocardium
10. Papillary muscle
11. Left ventricle

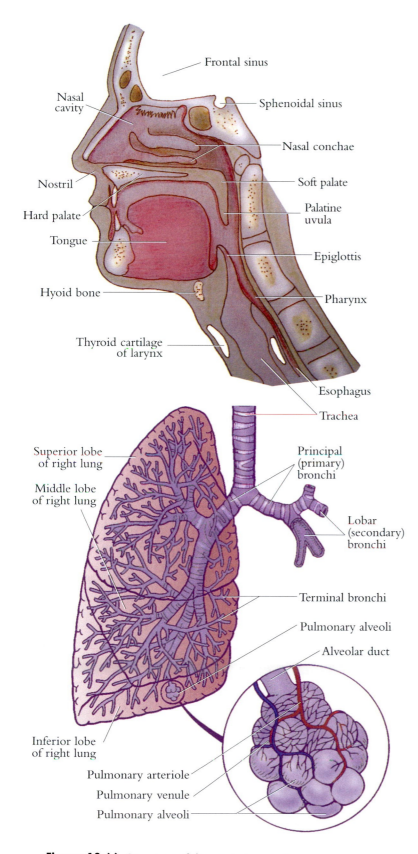

Frontal sinus

Nasal cavity

Sphenoidal sinus

Nasal conchae

Nostril

Soft palate

Hard palate

Palatine uvula

Tongue

Epiglottis

Hyoid bone

Pharynx

Thyroid cartilage of larynx

Esophagus

Trachea

Superior lobe of right lung

Principal (primary) bronchi

Middle lobe of right lung

Lobar (secondary) bronchi

Inferior lobe of right lung

Terminal bronchi

Pulmonary alveoli

Alveolar duct

Pulmonary arteriole

Pulmonary venule

Pulmonary alveoli

Figure 18.66 Structure of the respiratory system.

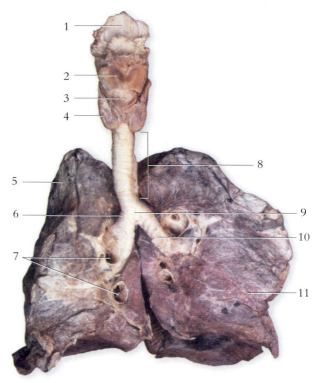

Figure 18.67 Anterior view of the larynx, trachea, and lungs.

1. Epiglottis
2. Thyroid cartilage
3. Cricoid cartilage
4. Thyroid gland
5. Right lung
6. Right principal (primary) bronchus
7. Pulmonary vessels
8. Trachea
9. Carnia
10. Left principal (primary) bronchus
11. Left lung

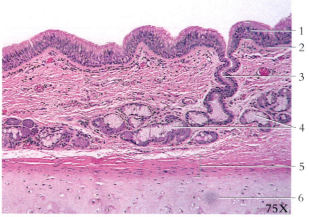

Figure 18.68 Tracheal wall.

1. Respiratory epithelium
2. Basement membrane
3. Duct of seromucous gland
4. Seromucous glands
5. Perichondrium
6. Hyaline cartilage

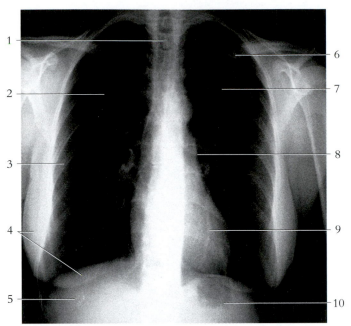

Figure 18.69 Radiograph of the thorax.
1. Thoracic vertebra
2. Right lung
3. Rib
4. Image of right breast
5. Diaphragm/liver
6. Clavicle
7. Left lung
8. Mediastinum
9. Heart
10. Diaphragm/stomach

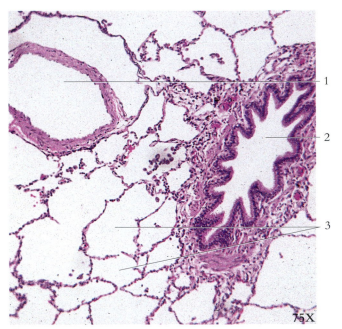

Figure 18.70 Bronchiole.
1. Pulmonary arteriole
2. Bronchiole
3. Pulmonary alveoli

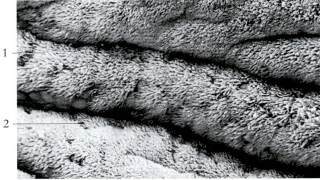

Figure 18.71 Electron micrograph of the lining of the trachea.
1. Cilia
2. Goblet cell

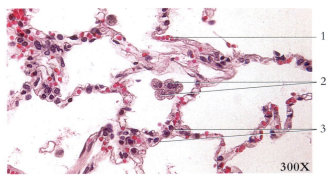

Figure 18.72 Pulmonary alveoli.
1. Capillary in alveolar wall
2. Macrophages
3. Type II pneumocytes

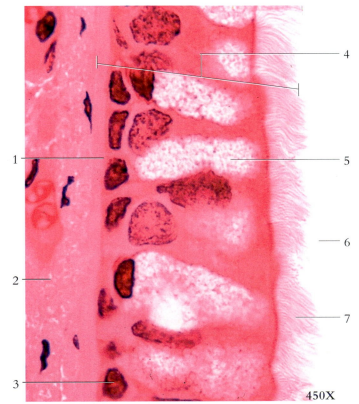

Figure 18.73 Bronchus.
1. Basement membrane
2. Lamina propria
3. Nucleus
4. Pseudostratified squamous epithelium
5. Goblet cell
6. Lumen of bronchus
7. Cilia

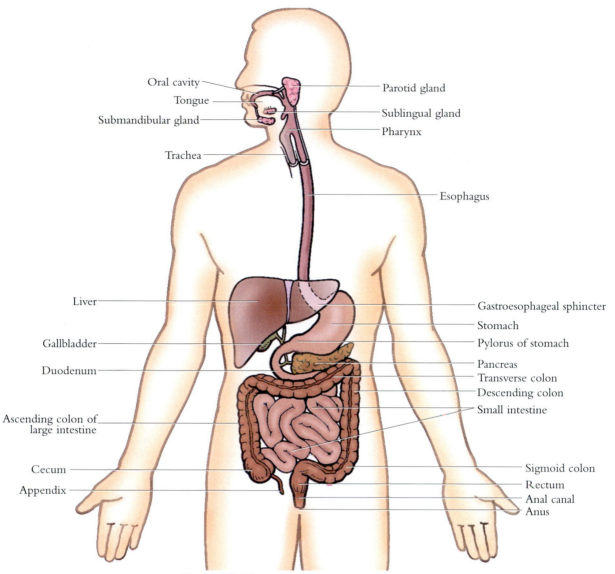

Figure 18.74 Structure of the digestive system.

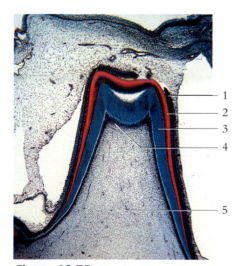

Figure 18.75 Developing tooth.
1. Ameloblasts   4. Odontoblasts
2. Enamel   5. Pulp
3. Dentin

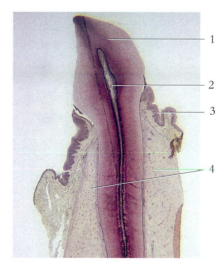

Figure 18.76 Mature tooth.
1. Dentin (enamel has   3. Gingiva
   been dissolved away)   4. Alveolar bone
2. Pulp

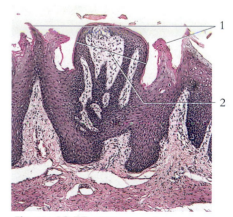

Figure 18.77 Filiform and fungiform papillae.
1. Filiform papillae
2. Fungiform papilla

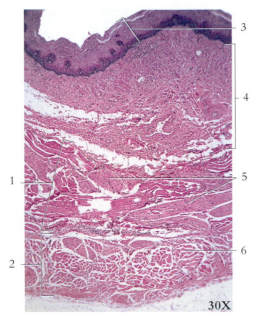

Figure 18.78 Wall of esophagus.
1. Inner circular layer (muscularis externa)
2. Outer longitudinal layer (muscularis externa)
3. Mucosa
4. Submucosa
5. Smooth muscle
6. Skeletal muscle

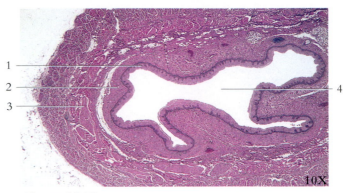

Figure 18.79 Transverse section of esophagus.
1. Mucosa
2. Submucosa
3. Muscularis
4. Lumen

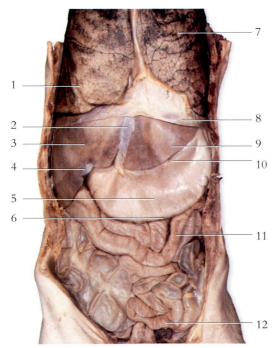

Figure 18.80 Anterior aspect of the trunk.
1. Right lung
2. Falciform ligament
3. Right lobe of liver
4. Gallbladder
5. Body of stomach
6. Greater curvature of stomach
7. Left lung
8. Diaphragm
9. Left lobe of liver
10. Lesser curvature of stomach
11. Transverse colon
12. Small intestine

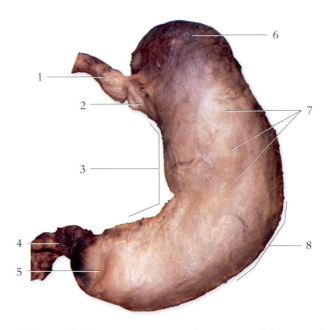

Figure 18.81 Major regions and structures of the stomach.
1. Esophagus
2. Cardiac portion of stomach
3. Lesser curvature of stomach
4. Duodenum
5. Pylorus of stomach
6. Fundus of stomach
7. Body of stomach
8. Greater curvature of stomach

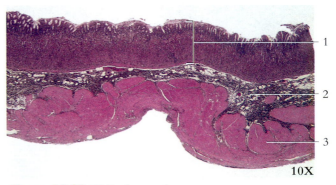

Figure 18.82 Wall of stomach.
1. Mucosa
2. Submucosa
3. Muscularis externa

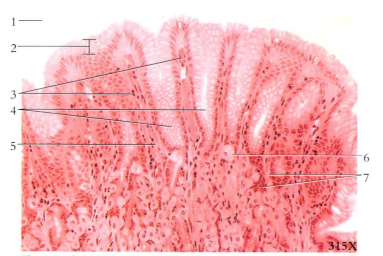

Figure 18.83   Histology of the cardiac region of the stomach.
1. Lumen of stomach
2. Surface epithelium
3. Mucosal ridges
4. Gastric pits
5. Lamina propria
6. Parietal cells
7. Chief (zymogenic) cells

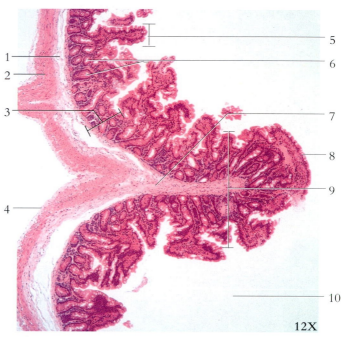

315X

Figure 18.84   Histology of the jejunum of the small intestine.
1. Submucosa
2. Circular and longitudinal muscles
3. Mucosa
4. Serosa
5. Villus
6. Intestinal glands
7. Submucosa
8. Microvilli
9. Plica circulares
10. Lumen of small intestine

12X

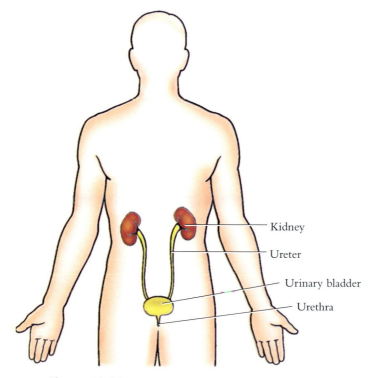

Kidney
Ureter
Urinary bladder
Urethra

Figure 18.85   Organs of the urinary system.

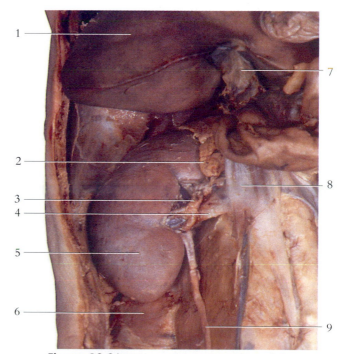

Figure 18.86   Kidney and ureter with overlying viscera removed.
1. Liver
2. Adrenal gland
3. Renal artery
4. Renal vein
5. Right kidney
6. Quadratus lumborum muscle
7. Gallbladder
8. Inferior vena cava
9. Ureter

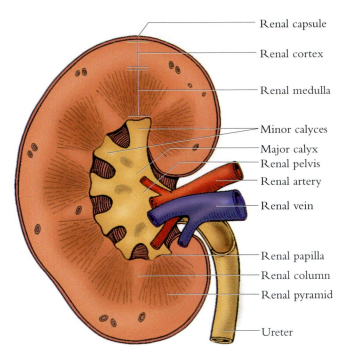

Renal capsule
Renal cortex
Renal medulla
Minor calyces
Major calyx
Renal pelvis
Renal artery
Renal vein
Renal papilla
Renal column
Renal pyramid
Ureter

Figure 18.87 Structure of the kidney.

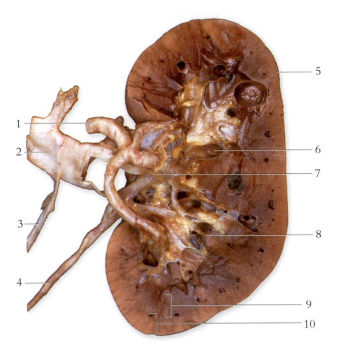

Figure 18.88 Coronal section of the left kidney.

1. Renal artery   6. Major calyx
2. Renal vein   7. Renal pelvis
3. Left testicular vein   8. Renal papilla
4. Ureter   9. Renal medulla
5. Renal capsule   10. Renal cortex

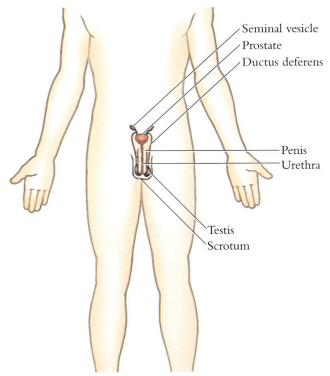

Seminal vesicle
Prostate
Ductus deferens
Penis
Urethra
Testis
Scrotum

Figure 18.89 Organs of the male reproductive system.

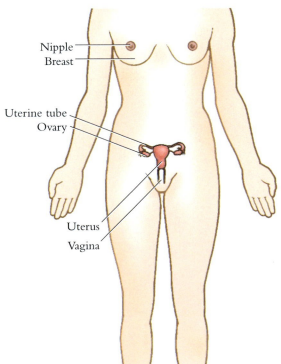

Nipple
Breast
Uterine tube
Ovary
Uterus
Vagina

Figure 18.90 Organs of the female reproductive system.

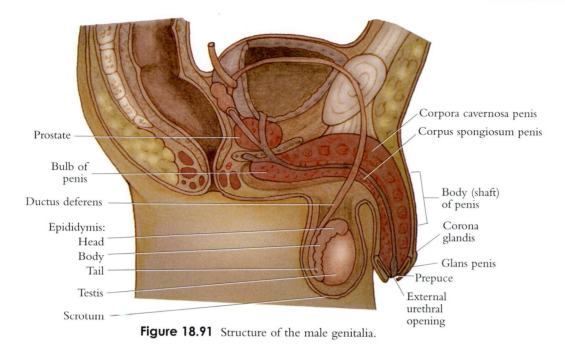

Prostate

Bulb of
penis

Ductus deferens

Epididymis:
Head
Body
Tail

Testis

Scrotum

Corpora cavernosa penis

Corpus spongiosum penis

Body (shaft)
of penis

Corona
glandis

Glans penis

Prepuce

External
urethral
opening

Figure 18.91 Structure of the male genitalia.

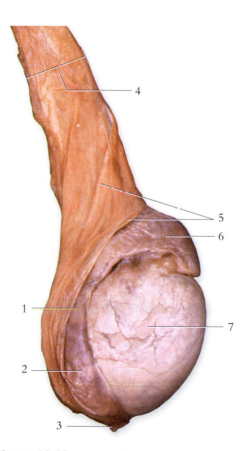

Figure 18.92 Testis and associated structures.
1. Body of epididymis   5. Spermatic fascia
2. Tail of epididymis   6. Head of epididymis
3. Gubernaculum   7. Testis
4. Spermatic cord

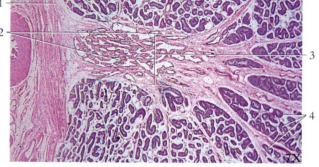

Figure 18.93 Testis.
1. Tunic albuginea   3. Mediastinum
2. Tubules of rete testis 4. Seminiferous tubules

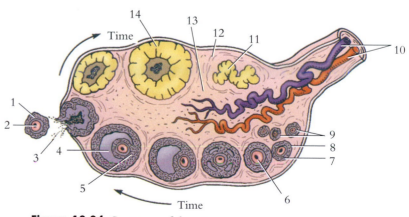

Time

Time

Figure 18.94 Structure of the ovary.
1. Corona radiata   8. Germinal epithelium
2. Secondary oocyte   9. Primary follicles
3. Ovulation   10. Ovarian vessels
4. Follicular fluid within antrum   11. Corpus albicans
5. Cumulus oophorus   12. Ovarian cortex
6. Oocyte   13. Ovarian medulla
7. Follicular cells   14. Corpus luteum

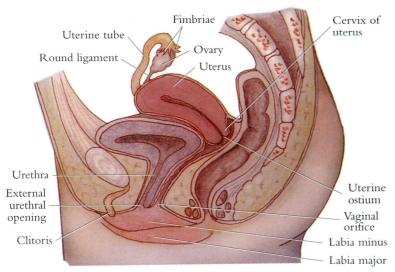

Figure 18.95 External genitalia and internal reproductive organs of the female reproductive system.

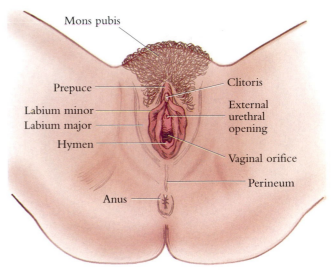

Figure 18.96 Female external genitalia.

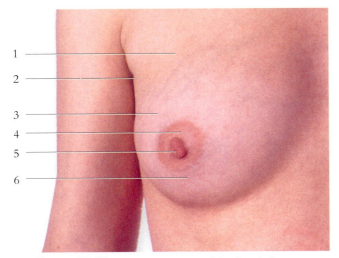

Figure 18.97 Surface anatomy of the female breast.
1. Pectoralis major muscle
2. Axilla
3. Lateral process of breast
4. Areola
5. Nipple
6. Breast (containing mammary glands)

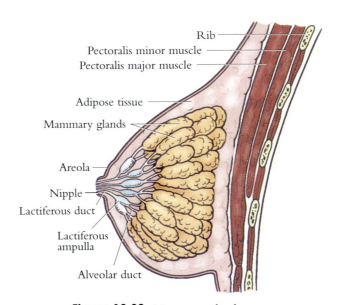

Figure 18.98 Mammary gland.

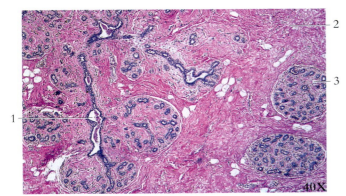

Figure 18.99 Mammary glands, (non-lactating glands).
1. Interlobular duct
2. Interlobular connective tissue
3. Lobule of glandular tissue

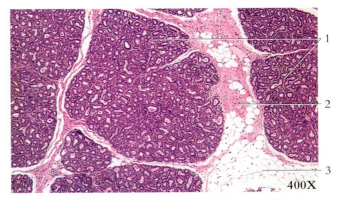

Figure 18.100 Mammary glands, (lactating glands).
1. Lobules of glandular tissue
2. Intralobular connective tissue
3. Adipose cells

Glossary of Terms

A

abdomen: the region of the mammalian body located between the diaphragm and the pelvis, that contains the abdominal cavity and its visceral organs; one of the three principal body regions (head, thorax, and abdomen) of many animals.

abduction: a movement away from the axis or midline of the body; opposite of adduction.

abiotic: without living organisms; nonliving portions of the environment.

acapnia: a decrease in normal amount of CO_2 in the blood.

accommodation: a change in the shape of the lens of the eye so that vision is more acute; the focusing for various distances.

acetone: an organic compound that may be present in the urine of diabetics; also called ketone bodies.

acetylcholine: a neurotransmitter chemical secreted at the terminal ends of many axons, responsible for postsynaptic transmission; also called ACh.

acetylcholinesterase: an enzyme that breaks down acetylcholine; also called AChE.

Achilles tendon: see *tendo calcaneus*.

acid: a substance that releases hydrogen ions (H^+) in a solution; a solution in which the pH is less than 7; acidic.

acidosis: a disorder of body chemistry in which the alkaline substances of the blood are reduced below normal.

acoelomate: without a coelomic cavity; as in flatworms.

acoustic: referring to sound or the sense of hearing.

actin: a protein in muscle fibers that together with myosin is responsible for contraction.

action potential: the change in ionic charge propagated along the membrane of a neuron; the nerve impulse.

active transport: movement of a substance into or out of a cell from a lesser to a greater concentration, requiring a carrier molecule and expenditure of energy.

adaptation: structural, physiological, or behavioral traits of an organism that promote its survival and contribute to its ability to reproduce under environmental conditions.

adduction: a movement toward the axis or midline of the body; opposite of abduction.

adenohypophysis: anterior pituitary gland.

adenoid: paired lymphoid structures in the naso-pharynx; also called pharyngeal tonsils.

adenosine triphosphate (ATP): a chemical compound that provides energy for cellular use.

adipose: fat, or fat-containing, such as adipose tissue.

adrenal glands: endocrine glands; one superior to each kidney; also called suprarenal glands.

aerobic: requiring free O_2 for growth and metabolism as in the case of certain bacteria called aerobes.

agglutination: clumping of cells; particular reference to red blood cells in an antigen antibody reaction.

aggression: provoking, domineering behavior.

allantois: an extraembryonic membranous sac that forms blood cells and gives rise to the fetal umbilical arteries and vein. It also contributes to the formation of the urinary bladder.

allele: an alternative form of a gene occurring at a given chromosome site, or locus.

all-or-none response: functioning completely when exposed to a stimulus of threshold strength; applies to action potentials through neurons and muscle fiber contraction.

alpha helix : right-handed spiral typical in proteins and DNA.

altruism: behavior benefiting other organisms without regard to its possible advantage or detrimental effect on the performer.

alveolus: A capsule within a structure. Pulmonary alveoli are the basic functional units of respiration.

amino acid: a unit of protein that contains an amino group (NH_2) and an acid group (COOH).

amnion: an extraembryonic membrane that surrounds the fetus to contain the amniotic fluid.

amniote: an animal that has an amnion during embryonic development; reptiles, birds, and mammals.

amoeba: protozoans that move by means of pseudopodia.

amphiarthrosis: a slightly moveable joint.

anaerobic respiration: metabolizing and growing in the absence of oxygen.

analogous: similar in function regardless of developmental origin; generally in reference to similar adaptations.

anatomical position: the position in human anatomy in which there is an erect body stance with the eyes directed forward, the arms at the sides, and the palms of the hands facing forward.

anatomy: the branch of science cocerned with the structure of the body and the relationship of its organs.

annulus: a ringlike segment, such as body rings on leeches.

antebrachium: the forearm.

antenna: a sensory appendage on many species of invertebrate animals.

anterior (ventral): toward the front; the opposite of posterior (dorsal).

anticodon: three ("a triplet") nucleotides in transfer RNA that pairs with a complementary codon (triplet) in messenger RNA.

antigen: a foreign material, usually a protein, that triggers the immune system to produce antibodies.

anus: the terminal end of the GI tract, opening of the anal canal.

aorta: the major systemic vessel of the arterial portion of the circulatory system, emerging from the left ventricle.

apocrine gland: a type of sweat gland that functions in evaporative cooling.

apopyle: opening of the radial canal into the spongocoel of sponges.

appeasement: submissive behavior, usually soliciting an end of aggression.

appendix: a short pouch that attaches to the cecum.

aqueous humor: the watery fluid that fills the anterior and posterior chambers of the eye.

arbor vitae: the branching arrangement of white matter within the cerebellum.

archaebacteria: Prokaryotic organisms that represent an early group of simple life forms, similar to bacteria but more closely related to eukaryotes.

archenteron: A principal cavity of an embryo during the gastrula stage. Lined with endoderm, the archenteron develops into the digestive tract.

areola: the pigmented ring around the nipple.

artery: a blood vessel that carries blood away from the heart.

articular cartilage: a hyaline cartlaginous covering over the articulating surface of bones of synovial joints.

ascending colon: the portion of the large intestine between the cecum and the hepatic flexure.

asexual: lacking distinct sexual organs and lacking the ability to produce gametes.

aster: minute rays of microtubules at the ends of the spindle apparatus in animal cells during cell division.

asymmetry: not symetrical.

atom: the smallest unit of an element that can exist and still have the properties of the element; collectively, atoms form molecules in a compound.

atomic number: the number of protons within the nucleus of an atom.

atomic weight: the number of protons together with the number of neutrons within the nucleus of an atom.

ATP: a compound of adenine, ribose, and three phosphates, two of which are high-energy phospates; it is the energy source for most cellular processes.

atrium: either of two superior chambers of the heart that receive venous blood.

atrophy: a wasting away or decrease in size of a cell or organ.

auditory tube: a narrow canal that connects the middle-ear chamber to the pharynx; also called the eustachian canal.

autonomic: self-governing; pertaining to the division of the nervous system which controls involuntary activities.

autosome: a chromosome other than a sex chromosome.

autotroph: an organism capable of synthesizing its own organic molecules (food) from inorganic molecules.

axilla: the depressed hollow under the arm; the armpit.

axon: The elongated process of a neuron (nerve cell) that transmits an impulse away from the cell body.

B

bacteria: prokaryotes within the kingdom Monera, lacking the organelles of eukaryotic cells.

base: a substance that contributes or liberates hydroxide ions in a solution; a solution in which the pH is greater than 7; alkaline.

basement membrane: a thin sheet of extracellular substance to which the basal surfaces of epithelial cells are attached.

basophil: a granular leukocyte that readily stains with basophilic dye.

belly: the thickest circumference of a skeletal muscle.

benign: nonmalignant; a confined tumor.

bilateral symmetry: the morphologic condition of having similar right and left halves.

binary fission: a process of reproduction that does not involve a mitotic spindle.

binomial system: assignment of two names to an organism, the first of which is the genus and the second the specific epithet, together constituting the species name.

biome: a major climax community characterized by a particular group of plants and animals.

biosphere: the portion of the earth's atmosphere and surface where living organisms exist.

biotic: pertaining to aspects of life, especially to characteristics of populations or ecosystems.

blastocoel: the cavity of a blastocyst.

blastula: an early stage of prenatal development between the morula and embryonic stages.

blood: the fluid connective tissue that circulates through the cardiovascular system to transport substances throughout the body.

bolus: a moistened mass of food that is swallowed from the oral cavity into the pharynx.

bone: an organ composed of solid, rigid connective tissue, forming a component of the skeletal system.

Bowman's capsule: see *glomerular capsule*.

brain: the enlarged superior portion of the central nervous system, located in the cranial cavity of the skull.

brain stem: the portion of the brain consisting of the medulla oblongata, pons, and midbrain.

bronchial tree: the bronchi and their branching bronchioles.

bronchiole: a small division of a bronchus within the lung.

bronchus: a branch of the trachea that leads to a lung.

buccal cavity: the mouth, or oral cavity.

buffer: a compound or substance that prevents large changes in the pH of a solution.

bursa: a saclike structure filled with synovial fluid, which occurs around joints.

buttock: the rump or fleshy mass on the posterior aspect of the lower trunk, formed primarily by the gluteal muscles.

C

calorie: the heat required to raise one kilogram of water one degree centigrade.

calyx: a cup-shaped portion of the renal pelvis that encircles renal papillae.

cancellous bone: spongy bone; bone tissue with a latticelike structure.

capillary: a microscopic blood vessel that connects an arteriole and a venule; the functional unit of the circulatory system.

carapace: protective covering over the dorsal part of the body of certain crustaceans and turtles.

carcinogenic: stimulating or causing the growth of a malignant tumor, or cancer.

carnivore: any animal that feeds upon another; specifically, flesh-eating mammal.

carpus: pertaining to the wrist; the eight human wrist bones.

carrying capacity: the maximum number of organisms of a species that can be maintained indefinitely in an ecosystem.

cartilage: a type of connective tissue with a solid flexible matrix.

caudal: referring to a position more toward the tail.

cecum: the pouchlike portion of the large intestine to which the ileum of the small intestine is attached; also spelled caecum.

cell: the structural and functional unit of an organism; the smallest structure capable of performing all the functions necessary for life.

cellular respiration: the reactions of glycolysis, Krebs cycle, and electron transport system that provide cellular energy and accompanying reactions to produce ATP.

central nervous system (CNS): the brain and the spinal cord.

centriole: an organelle usually located in the centrosome, considered to be the active division center of the animal cell.

centromere: a portion of the chromosome to which a spindle fiber attaches during mitosis or meiosis.

centrosome: a dense body near the nucleus of a cell that contains a pair of centrioles.

cephalothorax: fusion of the head and thoracic regions, characteristics of certain arthropods.

cercaria: a larva of trematodes (flukes).

cerebellum: the portion of the brain concerned with the coordination of movements and equilibrium.

cerebrospinal fluid: a fluid that buoys and cushions the central nervous system.

cerebrum: the largest portion of the brain, composed of the right and left hemispheres.

cervical: pertaining to the neck or a necklike portion of an organ.

chelipeds: pairs of pincerlike legs in most decapod crustaceans, adapted for seizing and crushing.

chitin: strong, flexible polysaccharide forming the exoskeleton of arthropods.

cholesterol: a lipid used in the synthesis of steroid hormones.

chondrocyte: a cartilage cell.

chorion: An extraembryonic membrane that participates in the formation of the placenta.

choroid: the vascular, pigmented middle layer of the wall of the eye.

chromatin: threadlike network of DNA and proteins within the nucleus.

chromosome: structure in the nucleus that contains the genes for genetic expression.

chyme: The mass of partially digested food that passes from the stomach into the duodenum of the small intestine.

cilia: microscopic, hairlike processes that move in an oar like manner on the exposed surfaces of certain epithelial cells.

ciliary body: a portion of the choroid layer of the eye that secretes aqueous humor and contains the ciliary muscle.

ciliates: protozoans that move by means of cilia.

circadian rhythm: a daily physiological or behavioral event, occurring on an approximate 24 hour cycle.

circumduction: a conelike movement of a body part, such that the distal end moves in a circle while the proximal portion remains relatively stable.

clitoris: a small, erectile structure in the vulva of the female.

cloaca: terminal portion of the digestive tract of many animals, that also may serve the excretory, reproductive, and respiratory systems.

cochlea: the spiral portion of the inner ear that contains the spiral organ (organ of Corti).

clone: asexually produced organisms having consistent genetic constitution.

cnidarian: small aquatic organisms having radial symmetry and stinging cells with nematocysts.

cocoon: protective, or resting, stage of development in certain invertebrate animals.

codon: a "triplet" of three nucleotides in mRNA that directs the placement of an amino acid into a polypeptide chain.

coelom: a fluid-filed space lined with peritoneum in visceral cavity of many bilateral animals.

collar cells: flagella-supporting cells in the inner layer of the wall of sponges.

colon: the large intestine.

common bile duct: a tube that is formed by the union of the hepatic duct and cystic duct, transports bile to the duodenum.

compact bone: tightly packed bone that is superficial to spongy bone; also called dense bone.

competition: interaction between individuals of the same or different species for a mutually necessary resource.

compound eye: arthropod eye consisting of multiple lenses.

condyle: a rounded process at the end of a long bone that forms an articulation.

conjugation: sexual union in which the nuclear material of one cell enters another.

connective tissue: one of the four basic tissue types within the body. It is a binding and supportive tissue with abundant matrix.

consumer: an organism that derives nutrients by feeding upon another or it remains.

control: a sample in an experiment that undergoes all the steps in the experiment except the one being investigated.

coral: a cnidarian that has a calcium carbonate skeleton whose remains contribute to form reefs.

cornea: the transparent convex, anterior portion of the outer layer of the eye.

cortex: the outer layer of an organ such as the convoluted cerebrum, adrenal gland, or kidney.

costal cartilage: the cartilage that connects the ribs to the sternum.

cranial: pertaining to the cranium.

cranial nerve: one of twelve pairs of

nerves that arise from the inferior surface of the brain.

cranium: the bones of the skull that enclose the brain and support the organs of sight, hearing, and balance.

crossing over: the exchange of corresponding chromatid segments of genetic material of homologous chromosomes during synapsis of meiosis I.

cyanobacteria: photosynthetic prokaryotes that have chlorophyll and release oxygen.

cytokinesis: division of the cellular cytoplasm.

cytology: the science dealing with the study of cells.

cytoplasm: the protoplasm of a cell located outside of the nucleus.

cytoskeleton: protein filaments throughout the cytoplasm of certain cells that help maintain the cell shape.

D

dendrite: a nerve cell process that transmits impulses toward a neuron cell body.

denitrifying bacteria: single-cellular organisms that convert nitrate to atmospheric nitrogen.

dentin: the principal substance of a tooth, covered by enamel over the crown and by cementum on the root.

dermis: the second, or deep, layer of skin beneath the epidermis.

descending colon: the segment of the large intestine that descends on the left side from the level of the spleen to the level of the left iliac crest.

diaphragm: a flat dome of muscle and connective tissue that separates the thoracic and abdominal cavities in mammals.

diaphysis: the shaft of a long bone.

diastole: the sequence of the cardiac cycle during which the ventricular heart chamber wall is relaxed.

diarthrosis: a freely movable joint.

diffusion: movement of molecules from an area of greater concentration to an area of lesser concentration.

dihybrid cross: a breeding experiment in which parental varieties differing in two traits are mated.

dimorphism: occurrence of two distinct forms within a species, with regards to size, color, organ structure, and so on.

diphyodont: two sets of teeth, deciduous and permanent.

diploid: twice (2n) the number of chromosomes found in gametes.

distal: away from the midline or origin; the opposite of proximal.

dominant: a hereditary characteristic that expresses itself even when the genotype is heterozygous.

dorsal: pertaining to the back or posterior portion of a body part; the opposite of ventral.

double helix: a double spiral used to describe the three-dimensional shape of DNA.

ductus deferens: a tube that carries spermatozoa from the epididymis to the ejaculatory duct: also called the vas deferens or seminal duct.

duodenum: the first portion of the small intestine.

dura mater: the outermost meninx (fibrous membrane) covering the central nervous system.

E

eccrine gland: a sweat gland that functions in body cooling; secrets pheromones.

ecology: the study of the relationship of organisms between themselves and the physical environment.

ecosystem: a biological community and its associated abiotic environment.

ectoderm: the outermost of the three primary, embryonic germ layers that gives rise to skin and nervous tissue.

edema: an excessive retention of fluid in tissues.

effector: an organ such as a gland or muscle that responds to motor stimulation.

efferent: conveying away from the center of an organ or structure.

ejaculation: the discharge of semen from the male urethra during climax.

electrocardiogram: a recording of the electrical activity that accompanies the cardiac cycle; also called ECG or EKG.

electroencephalogram: a recording of the brain wave pattern; also called EEG.

electromyogram: a recording of the activity of a muscle during contraction: also called EMG.

electrolyte: a solution that conducts electricity by means of charged ions.

electron: the unit of negative electricity.

element: a structure comprised of only one type of atom (i.e., carbon, hydrogen, oxygen).

emulsification: the process of dispersing one liquid in another.

enamel: the outer, dense substance covering the crown of a tooth.

endocardium: the fibrous lining of the heart chambers and valves.

endochondral bone: bones that form as hyaline cartilage models first and then are ossified.

endocrine gland: a hormone-producing gland that secretes directly into the blood or body fluids.

endoderm: the innermost of the three primary germ layers of an embryo that gives rise to digestive system.

endometrium: the inner lining of the uterus.

endoskeleton: hardened, supportive internal tissue of echinoderms and vertebrates.

endothelium: the layer of epithelial tissue that forms the thin inner lining of blood vessels and heart chambers.

enzyme: a protein catalyst that activates a specific reaction.

eosinophil: a type of white blood cell that becomes stained by acidic eosin dye; constitutes about 2%–4% of the human white blood cells.

epicardium: the thin, outer layer of the heart: also called the visceral pericardium.

epidermis: the outermost layer of the skin, composed of stratified squamous epithelium.

epididymis: a coiled tube located along the posterior border of the testis; stores spermatozoa and discharges them during ejaculation.

epidural space: a space between the spinal dura mater and the bone of the vertebral canal.

epiglottis: a cartilaginous leaf-like structure positioned on top of the larynx that covers the glottis during swallowing in mammals.

epinephrine: a hormone secreted from the adrenal medulla resulting in actions similar to those from sympathetic nervous system stimulation; also called adrenaline.

epiphyseal plate: a cartilaginous layer located between the epiphysis and diaphysis of a long bone. Functions in longitudinal bone growth.

epiphysis: the end segment of a long bone, distinct in early life but later becoming part of the larger bone.

epithelial tissue: one of the four basic tissue types; the type of tissue that covers or lines all exposed body surfaces.

erection: a response within an organ, such as the penis, when it becomes turgid and erect as opposed to being flaccid.

erythrocyte: a red blood cell.

esophagus: a tubular organ of the GI tract that leads from the pharynx to the stomach.

estrogen: female sex hormone secreted from the ovarian (Graafian) follicle.

estuary: a zone of mixing between fresh water and sea-water.

eukaryotic: possessing a nucleus and other membranous organelles characteristic of complex cells.

eustachian canal: see *auditory tube*.

evolution: organic evolution is any genetic change in organisms over time, or more precisely a change in gene frequency from one generation to another.

excretion: discharging waste material.

exocrine gland: a gland that secretes its product to an epithelial surface, directly or through ducts.

exoskeleton: an outer, hardened supporting structure secreted by ectoderm or epidermis.

expiration: the process of expelling air from the lungs through breathing out; also called exhalation.

extension: a movement that increases the angle between two bones of a joint.

external ear: the outer portion of the ear, consisting of the auricle (pinna), external auditory canal, and tympanum.

external nares: the opening into the nasal cavity; also called nostrils.

extracellular: outside a cell or cells.

extraembryonic membranes: membranes that are not a part of the embryo but are essential for the health and development of the organism.

extrinsic: pertaining to an outside or external origin.

F

facet: a small, smooth surface of a bone where articulation occurs.

facilitated transport: transfer of a particle into or out of a cell along a concentration gradient by a process requiring a carrier.

fallopian tube: see *uterine tube*.

false vocal cords: the supporting folds of tissue for the true vocal cords within the larynx.

fascia: a tough sheet of fibrous connective tissue binding the skin to underlying muscles or supporting and separating muscle.

fasciculus: a bundle of muscle or nerve fibers.

feces: waste material expelled from the GI tract during defecation, composed of food residue, bacteria, and secretions; also called stool.

fetus: the unborn offspring during the last stage of prenatal development.

filter feeder: an animal that obtains food by straining it from the water.

filtration: the passage of a liquid through a filter or a membrane.

fimbriae: fringelike extensions from the open end of the uterine tube.

fissure: a groove or narrow cleft that separates two parts of an organ.

flagella: long slender locomotor processes characteristic of flagellate protozoans, certain bacteria, and sperm.

flexion: a movement that decreases the angle between two bones of a joint; opposite of extension.

fluke: a parasitic flatworm within the Class Trematoda.

follicle: the portion of the ovary that produces the egg and the female sex hormone, estrogen; the depression that supports and develops a feather or hair.

fontanel: a membranous-covered region on the skull of a fetus or baby where ossification has not yet occurred: also called a soft spot.

food web: the food links between population in a community.

foot: the terminal portion of the lower extremity, consisting of the tarsus, metatarsus, and phalanges; a supporting structure used for locomotion.

foramen: an opening in a bone for the passage of a blood vessel or a nerve.

foramen ovale: the opening through the interatrial septum of the fetal heart.

fossa: a depressed area, usually on a bone.

fossil: any preserved remains or impressions of an organism within the earth's crust.

fourth ventricle: a cavity within the brain containing cerebrospinal fluid.

fovea centralis: a depression on the macula lutea of the eye where only cones are located, which is the area of keenest vision.

G

gallbladder: a pouchlike organ, attached to the inferior side of the liver, which stores and concentrates bile.

gamete: a haploid sex cell, sperm or egg.

gamma globulins: protein substances often found in immune serums that act as antibodies.

ganglion: an aggregation of nerve cell bodies outside the central nervous system.

gastrointestinal tract: the tubular portion of the digestive system that includes the stomach and the small and large intestines; also called the GI tract or alimentary canal.

gene: one of the biologic units of heredity; parts of the DNA molecule located in a definite position on a certain chromosome.

gene pool: the total of all the genes of the individuals in a population.

genetic drift: evolution by chance process.

genetics : the study of heredity.

genotype: the genetic makeup of an organism.

genus: the taxonomic category above species and below family.

gill: a gas exchange organ characteristic of fishes and other aquatic or semiaquatic animals.

gingiva: the fleshy covering over the mandible and maxilla through which the teeth protrude within the mouth; also called the gum.

gland: an organ that produces a specific substance or secretion.

glans penis: the enlarged, distal end of the penis.

glomerular capsule: the double-walled proximal portion of a renal tubule that encloses the glomerulus of a nephron; also called Bowman's capsule.

glomerulus: a coiled tuft of capillaries that is surrounded by the glomerular capsule and filters urine from the blood.

glottis: a slitlike opening into the larynx, positioned between the true vocal cords.

glycogen: the principal storage carbohydrate in animals. It is stored primarily in the liver and is made available as glucose when needed by the body cells.

goblet cell: a unicellular gland within columnar epithelia that secretes mucus.

gonad: a reproductive organ, testis or ovary, that produces gametes and sex hormones.

gray matter: the portion of the central nervous system that is composed of nonmyelinated nervous tissue.

grazer: animals that feed on low growing vegetation, such as grasses.

greater omentum: a double-layered serous membrane that originates on the greater curvature of the stomach and extends over the abdominal viscera.

gut: pertaining to the intestine; generally a developmental term.

gyrus: a convoluted elevation or ridge.

H

habitat: the ecological abode of a plant or animal species.

hair: an epidermal structure consisting of keratinized dead cells that have been pushed up from a dividing basal layer.

hair cells: specialized receptor nerve endings for responding to sensations, such as in the spiral organ of the inner ear.

hair follicle: a tubular depression in the skin in which a hair develops.

hand: the terminal portion of the upper extremity, consisting of the carpus, metacarpus, and phalanges.

haploid: the number (n) of unpaired chromosomes.

hard palate: the bony partition between the oral and nasal cavities, formed by the maxillae and palatine bones.

haustra: sacculations or pouches of the colon.

haversian system: see *osteon*.

heart: a muscular, pumping organ positioned in the thoracic cavity.

hematocrit: the volume percentage of red blood cells in whole blood.

hemoglobin: the pigment of red blood cells that transports O_2 and CO_2.

hemopoiesis: production of red blood cells.

hepatic portal circulation: the return of venous blood from the digestive organs through a capillary network within the liver before draining into the heart.

herbivore: an organism that feeds exclusively on plants.

heredity: the transmission of certain characteristics, or traits, from parents to offspring, via the genes.

heterodont: having teeth differentiated into incisors, canines, premolars, and molars for specific functions.

heterotroph: an organism that utilizes preformed food.

heterozygous: having two different alleles (i.e., Bb) for a given trait.

hiatus: an opening or fissure.

hilum: a concave or depressed area where vessels or nerves enter or exit an organ.

histology: microscopic anatomy of the structure and function of tissues.

homeostasis: self-regulation of body functions to maintain an internal steady state.

homologous: similar in developmental origin and sharing a common ancestry.

hormone: a chemical substance that is produced in an endocrine gland and secreted into the bloodstream to cause an effect in a specific target organ.

host: an organism on or in which another organism lives.

hyaline cartilage: the most common kind of cartilage in the body, occurring at the articular ends of bones, in the trachea, and within the nose, and forms the precursor to most of the bones of the skeleton.

hybrid: an offspring from the crossing of genetically different strains or species.

hymen: a developmental remnant of membranous tissue that partially covers the vaginal opening.

hyperextension: extension beyond the normal anatomical position of 180°.

hypothalamus: a structure within the brain below the thalamus, which functions as an autonomic center and regulates the pituitary gland.

hypothesis: a theory that is capable of explaining data and that may be used to predict the outcome of future experimentation.

hypotonic solution: a fluid environment that has a greater concentration of water and a lesser concentration of solute than the cell.

I

ileocecal valve: a specialization of the mucosa at the junction of the small and large intestine that forms a one-way passage and prevents the backflow of food materials.

ileum: the terminal portion of the small intestine between the jejunum and cecum.

imprinting: a type of learned behavior during a limited critical period.

indigenous: organisms that are native to a particular region; not introduced.

inferior vena cava: a systemic vein that collects blood from the body regions inferior (posterior) to the level of the heart and returns it to the right atrium.

inguinal: pertaining to the groin region.

inguinal canal: the circular passage through which a testis descends into the scrotum.

insertion: the more movable attachment of a muscle, usually more distal in location.

inspiration: the act of breathing air into the alveoli of the lungs; also called inhalation.

instar: stage of insect or other arthropod development between molts.

integument: pertaining to the skin.

internal ear: the innermost portion or chamber of the ear, containing the cochlea and the vestibular organs.

internal nares: the two posterior openings from the nasal cavity into the nasopharynx; also called the choanae.

interstitial: pertaining to spaces or structures between the functioning active tissue of any organ.

intracellular: within the cell itself.

intervertebral disc: a pad of fibrocartilage between the bodies of adjacent vertebrae.

intestinal gland: a simple tubular digestive gland that opens onto the surface of the intestinal mucosa and secretes digestive enzymes; also called crypt of Lieberkuhn.

intrinsic: situated or pertaining to internal origin.

invertebrate: an animal that lacks a vertebral column.

iris: the pigmented muscular portion of the eye that surounds the pupil and regulates its diameter.

islets of Langerhans: see *pancreatic islets*.

isotope: a chemical element that has the same atomic number as another but a different atomic weight.

isthmus: a narrow neck or portion of tissue connecting two structures.

J

jejunum: the middle portion of the small intestine, located between the duodenum and the ileum.

joint capsule: a fibrous tissue cuff surrounding a movable joint.

jugular: pertaining to the veins of the neck which drain the areas supplied by the carotid arteries.

K

karyotype: the arrangement of chromosomes that is characteristic of the species or of a certain individual.

keratin: an insoluble protein present in the epidermis and in epidermal derivatives such as scales, feathers, hair, and nails.

kidney: one of the paired organs of the urinary system that contains nephrons and filters urine from the blood.

L

labia majora: a portion of the external genitalia of a female, consisting of two longitudinal folds of skin extending downward and backward from the mons pubis.

labia minora: two small folds of skin, devoid of hair and sweat glands, lying between the labia majora of the external genitalia of a female.

lacrimal gland: a tear-secreting gland, located on the superior lateral portion of the eyeball underneath the upper eyelid.

lactation: the production and secretion of milk by the mammary glands.

lacteal: a small lymphatic duct within a villus of the small intestine.

lacuna: a hollow chamber that houses an osteocyte in mature bone tissue or a chondrocyte in cartilage tissue.

lamella: a concentric ring of matrix surrounding the central canal in an osteon of mature bone tissue.

large intestine: the last major portion of the GI tract, consisting of the cecum, colon, rectum, and anal canal.

larva: an immature, developmental stage that is quite different from the adult.

larynx: the structure located between the pharynx and trachea that houses the vocal cords; commonly called the voice box.

lens: a transparent refractive structure of the eye, derived from ectoderm and positioned posterior to the pupil and iris.

leukocyte: a white blood cell; also spelled leucocyte.

ligament: a fibrous band of connective tissue that binds bone to bone to strengthen and provide support to the joint; also may support viscera.

limbic system: a portion of the brain concerned with emotions and autonomic activity.

linea alba: a fibrous band extending down the anterior medial portion of the abdominal wall.

locus: the specific location or site of a gene within the chromosome.

lumbar: pertaining to the region of the loins.

lumen: the space within a tubular structure through which a substance passes.

lung: one of the two major organs of respiration within the thoracic cavity.

lymph: a clear fluid that flows through lymphatic vessels.

lymph node: a small, oval mass located along the course of lymph vessels.

lymphocyte: a type of white blood cell characterized by a granular cytoplasm.

M

macula lutea: a depression in the retina that contains the fovea centralis, the area of keenest vision.

malignant: a disorder that becomes worse and eventually causes death, as in cancer.

malnutrition: any abnormal assimilation of food; receiving insufficient nutrients.

mammary gland: the gland of the mammalian female breast responsible for lactation and nourishment of the young.

mantle: fleshy fold that envelops the viscera of a mollusk and secretes the shell.

marrow: the soft vascular tissue that occupies the inner cavity of certain bones and produces blood cells.

matrix: the intercellular substance of a tissue.

meatus: an opening or passageway into a structure.

mediastinum: the space in the center of the thorax between the two pleural cavities.

medulla: the center portion of an organ.

medulla oblongata: a portion of the brain stem between the pons and the spinal cord.

medullary cavity: the hollow center of the diaphysis of a long bone, occupied by marrow.

meiosis: nuclear reduction division by which gametes, or haploid sex cells, are formed after cytokinesis.

melanocyte: a pigment-producing cell in the deepest epidermal layer of the skin.

membranous bone: bone that forms from membranous connective tissue rather than from cartilage.

menarche: the first menstrual discharge in the human female.

meninges: a group of three fibrous membranes that cover the central nervous system.

menisci: wedge-shaped cartilage in certain movable joints.

menopause: the cessation of menstrual periods in the human female.

menses: the monthly flow of blood from the human female genital tract.

menstrual cycle: the rhythmic female reproductive cycle, characterized by changes in hormone levels and physical changes in the uterine lining.

menstruation: the discharge of blood and tissue from the uterus at the end of the human female menstrual cycle.

mesentery: a fold of peritoneal membrane that attaches an abdominal organ to the abdominal wall.

mesoderm: the middle of the three germ layers that gives rise to bone, muscle, blood, etc.

mesothelium: a simple squamous epithelial tissue that lines body cavities and covers visceral organs; also called serosa.

metabolism: the chemical changes that occur within a cell.

metacarpus: the region of the hand between the wrist and the phalanges, including the five bones that compose the palm of the hand.

metamorphosis: change in morphologic form, such as when an insect larva develops into the adult or as a tadpole develops into an adult frog.

metastasis: the spread of a disease from one organ or body part to another.

metatarsus: the region of the foot between the ankle and the phalanges, comprised of five bones.

microbiology: the science dealing with microscopic organisms, including bacteria, fungi, viruses, and protozoa.

microvilli: microscopic, hairlike projections of cell membranes on certain epithelial cells.

midbrain: the portion of the brain between the pons and the forebrain.

middle ear: the middle of the three ear chambers, containing the three ear ossicles in mammals.

migration: movement of organisms from one geographical site to another.

mimicry: a protective resemblance of an organism to another.

mitosis: the process of nuclear division, in which the two daughter cells are identical and contain the same number of nuclet chromosomes, typically followed by cytokinesis to form two identical daughter cells.

mitral valve: the left atrioventricular heart valve; also called the bicuspid valve.

mixed nerve: a nerve containing both motor and sensory nerve fibers.

molecule: a minute mass of matter, composed of a combination of atoms that form a given chemical substance or compound.

molting: periodic shedding of an epidermal derived structure.

motor neuron: a nerve cell that conducts action potential away from the central nervous system and innervates effector organs (muscles and glands); also called efferent neuron.

motor unit: a single motor neuron and the muscle fibers it innervates.

mucosa: a mucous membrane that lines cavities and tracts opening to the exterior.

muscle: an organ adapted to contract; three types of muscle tissue are cardiac, smooth, and skeletal.

mutation: a variation in an inheritable characteristic, a permanent transmissible change in which the offspring differ from the parents.

mutualism: a beneficial relationship between two organisms of different species.

myelin: a lipoprotein material that forms a sheathlike covering around nerve fibers.

myocardium: the cardiac muscle layer of the heart.

myofibril: a bundle of contractile fibers within muscle cells.

myoneural junction: the site of contact between an axon of a motor neuron and a muscle fiber.

myosin: a thick filament protein that together with actin causes muscle contraction.

N

nail: a hardenend, keratinized plate that develops from the epidermis and forms a protective covering on the dorsal surfaces of the digits.

nasal cavity: a mucosa-lined space above the oral cavity, which is divided by a nasal septum and is the first chamber of the respiratory system.

nasal septum: a bony and cartilaginous partition that separates the nasal cavity into two portions.

natural selection: the evolutionary mechanism by which better adapted organisms are favored to reproduce and pass on their genes to the next generation.

nephron: the functional unit of the kidney, consisting of a glomerulus, glomerular capsule, convoluted tubules, and the loop of the nephron.

nerve: a bundle of nerve fibers outside the central nervous system.

neurofibril node: a gap in the myelin sheath of a nerve fiber; also called the node of Ranvier.

neuroglia: specialized supportive cells of the central nervous system.

neurolemmocyte: a specialized neuroglia cell that surrounds an axon fiber of a peripheral nerve and forms the neurilemmal sheath; also called the Schwann cell.

neuron: the structural and functional unit of the nervous system, composed of a cell body, dendrites, and an axon; also called a nerve cell.

neutron: a subatomic particle in the nucleus of an atom that has a weight of one atomic mass unit and carries no charge.

neutrophil: a type of phagocytic white blood cell.

niche: the position and functional role of an organism in its ecosystem.

nipple: a dark pigmented, rounded projection at the tip of the mammalian breast openings of mammary glands.

nitrogen fixation: a process carried out by certain organisms, such as by soil bacteria, whereby free atmospheric nitrogen is converted into ammonia or organic N.

node of Ranvier: see *neurofibril node*.

notochord: a flexible rod of connective tissue providing skeletal support for swimmimg muscles in certain chordates or their embryos. **nucleic acid**: an organic molecule composed of joined nucleotides, such as RNA and DNA.

nucleus: a spheroid membrane-bound body within a cell that contains the genetic factors of the cell.

nurse cells: specialized cells within the testes that supply nutrients to developing spermatozoa; also called sertoli cells.

O

olfactory: pertaining to the sense of smell.

oocyte: a developing egg cell.

oogenesis: the process of female gamete formation.

optic: pertaining to the eye and the sense of vision.

optic chiasma: an X-shaped structure on the inferior aspect of the brain where there is a partial crossing over of fibers in the optic nerves.

optic disc: a small region of the retina where the fibers of the ganglion neurons exit from the eyeball to form the optic nerve; also called the blind spot.

oral: pertaining to the mouth; also called buccal.

organ: a structure consisting of two or more tissues, which performs a specific function.

organelle: a minute structure of a cell with a specific function.

organism: an individual living creature.

orifice: an opening into a body cavity or tube.

origin: the place of muscle attachment onto the more stationary point or proximal bone; opposite of the insertion.

osmosis: the passage of a solvent, such as water, from a solution of lesser concentration to one of greater concentration through a semipermeable membrane.

ossicle: one of the three bones of the middle ear of mammals.

osteocyte: a mature bone cell.

osteon: a group of osteocytes and concentric lamellae surrounding a central canal within bone tissue; also called a haversian system.

oval window: see vestibular window

ovarian follicle: a developing ovum and its surrounding epithelial cells.

ovary: the female gonad in which ova and certain sexual hormones are produced.

oviduct: the tube that transports ova from the ovary to the uterus; also called the uterine tube or fallopian tube.

ovipositor: a structure at the posterior end of the abdomen in many female insects for laying eggs.

ovulation: the rupture of an ovarian follicle with the release of an ovum.

ovule: the female reproductive organ in a seed plant that contains the megasporangium where meiosis occurs and the female gametophyte is produced.

ovum: a secondary oocyte after ovulation but before fertilization.

P

palate: the roof of the mouth or oral cavity.

palmar: pertaining to the palm of the primate hand.

pancreas: organ in the abdominal cavity that secretes gastric juices into the GI tract and insulin and glucagon into the blood.

pancreatic islets: a cluster of cells within the pancreas that forms the endocrine portion of the pancreas; also called islets of Langerhans.

papillae: small nipple-like projections.

paranasal sinus: a mucous-lined air chamber that communicates with the nasal cavity.

parasite: an organism that resides in or on another from which it derives sustenance, but oofers no benefits to the host.

parasympathetic: pertaining to the division of the autonomic nervous system concerned with activities that restore and conserve metabolic energy.

parathyroids: small endocrine glands that are embedded on the posterior surface of the thyroid glands and are concerned with calcium metabolism.

parietal: pertaining to a wall of an organ or cavity.

parotid gland: one of the paired salivary glands on the sides of the face over the masseter muscle.

parturition: the process of childbirth.

pathogen: any disease-producing organism.

pectoral girdle: the portion of the skeleton that supports the anterior appendages in a vertebrate.

pelvic: pertaining to the pelvis.

pelvic girdle: the portion of the skeleton to which the posterior appendages are supported.

penis: the external male genital organ, through which urine passes during urination and which transports semen to the female during coitus.

pericardium: a protective serous membrane that surrounds the heart.

perineum: the floor of the pelvis, which is the region between the anus and the scrotum in the male and between the anus and the vulva in the female.

periosteum: a fibrous connective tissue covering the surface of bone.

peripheral nervous system: the nerves and ganglia of the nervous system that lie outside of the brain and spinal cord.

peristalsis: rhythmic contractions of smooth muscle in the walls of various tubular organs, which move the contents along.

peritoneum: the serous membrane that lines the abdominal cavity and covers the abdominal viscera; requisite lining at a true coelam.

phagocyte: any cell that engulfs other cells, including bacteria, or small foreign particles.

phalanx, pl. phalanges: a bone of the finger or toe.

pharynx: the region of the GI tract and respiratory system located at the back of the oral and nasal cavities and extending to the larynx anteriorly and the esophagus posteriorly; also called the throat.

phenotype: the appearance of an organism caused by the genotype and environmental influences.

pheromone: a chemical secreted by one organism that influences the behavior of another.

photoperiodism: the response of an organism to periods of light and dark.

photosynthesis: the process of using the energy of the sun to make carbohydrate from carbon dioxide and water.

physiology: the science that deals with the study of body functions.

pia mater: the innermost meninx that is in direct contact with the brain and spinal cord.

pineal gland: a small cone-shaped gland located in the roof of the third ventricle.

pituitary gland: a small, pea-shaped endocrine gland situated on the inferior surface of the brain that secretes a number of hormones; also called the hypophysis and commonly called the master gland.

placenta: the organ of metabolic exchange between the mother and the fetus.

plankton: aquatic, free-floating microscopic organisms.

plasma: the fluid, extracellular portion of circulating blood.

platelets: fragments of specific bone marrow cells that function in blood coagulation: also called thrombocytes.

pleural membranes: serous membranes that surround the lungs and line the thoracic cavity.

plexus: a network of interlaced nerves or vessels.

plica circulares: a deep fold within the wall of the small intestine that increases the absorptive surface area.

polypeptide: a molecule of many amino acids linked by peptide bonds.

pons: the portion of the brain stem just above the medulla oblongata and anterior to the cerebellum.

population: all the organisms of the same species in a particular location.

posterior (dorsal): toward the back or upper surface.

predation: the consumption of one organism by another.

pregnancy: a condition where a female has developing offspring in the uterus.

prenatal: the period of offspring development during pregnancy; before birth.

prey: organisms that are food for a predator.

producers: organisms that synthesize organic compounds from inorganic constituents within an ecosystem.

prokaryote: organism, such as a bacterium, that lacks an organized nucleus and specialized organelles.

proprioceptor: a sensory nerve ending that responds to changes in tension in a muscle or tendon.

prostate: a walnut-shaped gland surrounding the male urethra just below the urinary bladder that secretes an additive to seminal fluid during ejaculation.

protein: a macromolecule composed of one or several polypeptides.

proton: a subatomic particle of the atom nucleus that has a weight of one atomic mass unit and carries a positive charge; also a hydrogen ion.

proximal: closer to the midline of the body or origin of an appendage: opposite of distal.

puberty: the period of human development in which the reproductive organs become functional.

pulmonary: pertaining to the lungs.

pupil: the opening through the iris that permits light to enter the posterior cavity of the eyeball and be refracted by the lens.

R

receptor: a sense organ or a specialized end of a sensory neuron that receives stimuli from the environment.

rectum: the terminal portion of the GI tract, from the sigmoid colon to the anus.

reflex arc: the basic conduction pathway through the nervous system, consisting of a sensory neuron, interneuron, and a motor neuron.

regeneration: regrowth of tissue or the formation of a complete organism from a portion.

renal: pertaining to the kidney.

renal corpuscle: the portion of the nephron consisting of the glomerulus and a glomerular capsule.

renal pelvis: the inner cavity of the kidney formed by the expanded ureter and into which the calyces open.

renewable resource: a commodity that is continually produced in the environment.

replication: the process of producing a duplicate; a copying or duplication, such as DNA replication.

respiration: the exchange of gases between the external environment and the cells of an organism.

rete testis: a network of ducts in the center of the testis, site of spermatozoa production.

retina: the inner layer of the eye that contains the rods and cones.

retraction: the movement of a body part, such as the mandible, backward on a plane parallel with the ground; the opposite of protraction.

rod: a photoreceptor in the retina of the eye that is specialized for colorless, dim light vision.

rotation: the movement of a bone around its own longitudinal axis.

rugae: the folds or ridges of the mucosa of an organ.

S

sagittal: a vertical plane through the body that divides it into right and left portions.

salivary gland: an accessory digestive gland that secretes saliva into the oral cavity.

sarcolemma: the cell membrane of a muscle fiber.

sarcomere: the portion of a skeletal muscle fiber between the two adjacent Z lines that is considered the functional unit of a myofibril.

Schwann cell: see *neurolemmocyte*.

scientific method: process consisting of hypothesis generation, observation and experimentation, and results in testable theories; method by which reproducible data are obtained.

sclera: the outer white layer of connective tissue that forms the protective covering of the eye.

scolex: attachment region of a tapeworm.

scrotum: a pouch of skin that contains the testes and their accessory organs.

sebaceous gland: an exocrine gland of the skin that secretes sebum, an oily protective product.

semen: the secretion of the reproductive organs of the male, consisting of spermatozoa and additives.

semicircular canals: tubular channels within the inner ear that contain the receptors for equilibrium.

semilunar valve: crescent-shaped heart valves, positioned at the entrances to the aorta and the pulmonary trunk.

seminal vesicles: a pair of accessory male reproductive organs lying posterior and inferior to the urinary bladder, which secrete additives to spermatozoa into the ejaculatory ducts.

sensory neuron: a nerve cell that conducts an impulse from a receptor organ to the central nervous system; also called afferent neuron.

serous membrane: an epithelial and connective tissue membrane that lines body cavities and covers viscera; also called serosa.

sesamoid bone: a membranous bone formed in a tendon in response to joint stress.

sessile: organisms that lack locomotion and remain stationary, such as sponges and plants.

sigmoid colon: the S-shaped portion of the large intestine between the descending colon and the rectum.

sinoatrial node: a mass of cardiac tissue in the wall of the right atrium that initiates the cardiac cycle; the SA node; also called the pacemaker.

sinus: a cavity or hollow space within a body organ such as a bone.

skeletal muscle: a type of muscle tissue that is multinucleated, occurs in bundles, has crossbands of proteins, and contracts either in a typically voluntary or involuntary fashion.

small intestine: the portion of the GI tract between the stomach and the cecum, functions in absorption of food nutrients.

smooth muscle: a type of muscle tissue that is nonstriated, composed of fusiform, single-nucleated fibers, and contracts in an involuntary, rhythmic fashion within the walls of visceral organs.

solute: a substance dissolved in a solvent to form a solution.

solvent: a fluid such as water that dissolves solutes.

somatic: pertaining to the nonreproductive (nonvisceral) parts of the body.

species: a group of morphologically similar (common gene pool) organisms that are capable of interbreeding and producing fertile offspring.

spermatic cord: the structure of the male reproductive system composed of the ductus deferens, spermatic vessels, nerve, cremasteric muscle, and connective tissue.

spermatogenesis: the production of male sex gametes, or spermatozoa.

spermatozoon: a sperm cell, or gamete.

sphincter: a circular muscle that constricts a body opening or the lumen of a tubular structure.

spinal cord: the portion of the central nervous system that extends from the brain stem through the vertebral canal, also called "dorsal nerve cord".

spinal nerve: one of the thirty-one pairs of nerves that arise from the spinal cord.

spiracle : a respiratory opening in certain animals such as arthropods and sharks.

spleen: a large, blood-filled organ located in the upper left of the abdomen and attached by the mesenteries to the stomach.

spongy bone: a type of bone that contains many porous spaces; also called cancellous bone.

stomach: a pouchlike digestive organ between the esophagus and the duodenum.

submucosa: a layer of supportive connective tissue that underlies a mucous membrane.

succession: ecological stages by which a particular biotic community gradually changes until there is a community of climax vegetation.

superior vena cava: a large systemic vein that collects blood from regions of the body superior to the heart and returns it to the right atrium.

surfactant: a substance produced by the lungs that decreases the surface tension within the pulmonary alveoli.

suture: a type of immovable joint articulating between bones of the skull.

sympathetic: pertaining to that part of the autonomic nervous system concerned with processes involving the utilization of energy.

synapse: a minute space between the axon terminal of a presynaptic neuron and a dendrite of a postsynaptic neuron or an effecor cell.

synovial cavity: a space between the two bones of a diarthrotic joint, filled with synovial fluid.

system: a group of body organs that function together.

systole: the muscular contraction of the ventricles of the heart during the cardiac cycle.

systolic pressure: arterial blood pressure during the ventricular systolic phase of the cardiac cycle.

T

target organ: the specific body organ that a particular hormone affects.

tarsus: the seven bones that form the ankle.

taxonomy: the science of describing, classifying, and naming organisms.

tendo calcaneous: the tendon that attaches the calf muscles to the calcaneous bone.

tendon: a band of dense regular connective tissue that attaches muscle to bone.

tetrapod: a four-appendaged vertebrate, such as amphibian, reptile, bird, or mammal.

testis: the primary reproductive organ of a male, which produces spermatozoa and male sex hormones.

thoracic: pertaining to the chest region.

thoracic duct: the major lymphatic vessel of the body, which drains lymph from the entire body except the upper right quadrant and returns it to the left subclavian vein.

thorax: the chest.

thymus gland: a bi-lobed lymphoid organ positioned in the upper mediastinum, posterior to the sternum and between the lungs.

tissue: an aggregation of two or more types of cells and their binding intercellular substance, joined to perform a specific function.

tongue: a protrusible muscular organ on the floor of the oral cavity.

trachea: the airway leading from the larynx to the bronchi; also called the windpipe.

tract: a bundle of nerve fibers within the central nervous system.

trait: a distinguishing feature studied in heredity.

transverse colon: a portion of the large intestine that extends from right to left across the abdomen between the hepatic and splenic flexures.

tricuspid valve: the heart valve between the right atrium and the right ventricle.

true vocal cords: folds of the mucous membrane in the larynx that produce sound as they are pulled taut and vibrated.

turgor pressure: osmotic pressure that provides rigidity to a cell.

tympanic membrane: the membranous eardrum positioned between the outer and middle ear; also called the tympanum, or the ear drum.

U

umbilical cord: a cordlike structure containing the umbilical arteries and vein, which connects the fetus with the placenta.

umbilicus: the site where the umbilical cord was attached to the fetus: also called the navel.

ureter: a tube that transports urine from the kidney to the urinary bladder.

urethra: a tube that transports urine from the urinary bladder to the outside of the body.

urinary bladder: a distensible sac in the pelvic cavity which stores urine.

uterine tube: the tube through which the ovum is transported to the uterus and where fertilization takes place: also called the oviduct or fallopian tube.

uterus: a hollow, muscular organ in which a fetus develops. It is located within the female pelvis between the urinary bladder and the rectum.

uvula: a fleshy, pendulous portion of the soft palate that blocks the nasopharynx during swallowing.

V

vagina: a tubular organ that leads from the uterus to the vestibule of the female reproductive tract and receives the male penis during coitus.

vein: a blood vessel that conveys blood toward the heart.

ventral: toward the lower surface of the body.

vertebrate: an animal that possesses a vertebral column.

vestibular window: a membrane-covered opening in the bony wall between the middle and inner ear, into which the footplate of the stapes fits.

viscera: the organs within the abdominal or thoracic cavities.

vitreous humor: the transparent gell that occupies the space between the lens and retina of the eye.

vulva: the external genitalia of the female that surround the opening of the vagina; also called the pudendum.

Z

zygote: a fertilized egg cell formed by the union of a sperm and an ovum.

Index